**Basic Experiments
with Radioisotopes**

Basic Experiments with Radioisotopes

For courses in physics, chemistry and biology

~~~~~~

J N Andrews PhD, DIC, ARIC
Director of Nuclear Studies, University of Bath

D J Hornsey MSc, MIBiol
Lecturer in Radiobiology, University of Bath

SI Units

Pitman Publishing

First published 1972

SIR ISAAC PITMAN AND SONS LTD
Pitman House, Parker Street, Kingsway, London WC2B 5PB
PO Box 46038, Portal Street, Nairobi, Kenya
SIR ISAAC PITMAN (AUST) PTY LTD
Pitman House, 158 Bouverie Street, Carlton, Victoria 3053, Australia
PITMAN PUBLISHING COMPANY SA LTD
PO Box 11231, Johannesburg, South Africa
PITMAN PUBLISHING CORPORATION
6 East 43rd Street, New York, NY 10017, USA
SIR ISAAC PITMAN (CANADA) LTD
495 Wellington Street West, Toronto 135, Canada
THE COPP CLARK PUBLISHING COMPANY
517 Wellington Street West, Toronto 135, Canada

© J N Andrews and D J Hornsey 1972

ISBN 0 273 36152 X

Made in Great Britain at The Pitman Press, Bath
G2 (T333:78)

Preface

The application of radioisotope methods to research in the physical and biological sciences has become firmly established during the past two decades. Because of this a need has arisen for the provision of some basic instruction in the fundamentals of nuclear science for students in a wide variety of disciplines. Some aspects of the subject have been introduced into school curricula, initially through the interest of the Association for Science Education. In recent years the modernization of science teaching has been accelerated by the support of the Nuffield Foundation's Science Teaching Project, and work with radioisotopes has been included in several subjects at school level. In the field of further education, such introductory work with radioisotopes is generally developed to provide the student with a basic understanding of radioactivity and its applications in his discipline of study. It is against this background that the present volume was written with the aim of providing an essentially practical text which would serve as an introduction to the use of radioisotopes at school and university or technical college level. It is hoped that it will prove useful both as a teacher's reference for the introduction of radioactivity at school level and as an introduction to radioisotope work for the undergraduate or technical college student in physics, chemistry or biology. By including material from several scientific disciplines we hope we shall encourage readers to develop an awareness of the applications of radioisotopes in fields other than their own and perhaps even to carry out some work beyond the boundaries of their own discipline.

The background to nuclear physics, necessary for carrying out the practical work described, is included in the introductory chapters. No extensive previous knowledge of the subject has been assumed, but the treatment is not intended to be exhaustive, and suggestions for further reading have been included for those who wish to make a more detailed study. Some parts of these chapters have been set in smaller type and can be ignored, if desired, without affecting the development of the subject matter.

Radiation protection has been dealt with in some detail as it has been our experience that reluctance to use radioisotope methods is often due to a lack of knowledge concerning the extent of the radiation hazards involved. We have attempted to place these hazards in perspective.

The experimental work has been designed so that it involves mainly simple apparatus and very small amounts of radioactive materials. It has been our aim, so far as possible, to relate theoretical concepts to the practical work involved, and each experiment is generally preceded by a summary of the relevant theory. Some guidance is presented on page xiii for the selection of course work.

The units of the Système International d'Unités (SI) have been used throughout the book, together with some specialized units which do not belong to that system but are recognized by the Comité International des Poids et Mesures for use with it. These are the electronvolt and the unified atomic mass unit, which are accepted unconditionally; and the barn, curie, röntgen and rad, which are accepted "temporarily." We think it will be some considerable time before they are abandoned.

We should like to acknowledge with gratitude the help given to us in the preparation of the manuscript by Mr. G. E. Williams of Pitman Publishing.

J. N. A.
D. J. H.

Contents

	Preface	v
	Suggestions for course work	xiii
Chapter 1	Nuclei and radioactivity	1
2	Radioisotopes and labelled compounds	20
3	The detection of nuclear radiation	33
4	The biological basis of radiation protection	62
5	Radiation dosimetry and maximum permissible levels of dose	74
6	Practical methods of radiation protection	85
7	Regulations for the use of radioisotopes in educational establishments	98
Experiment 1	The determination of half-life	103
2	Radioactive decay phenomena simulated by the flow of water through tubes	107
3	Electroscopes and ion chambers for radiation detection	115
4	The Geiger–Müller counter	120
	4.1 Plateau curve and efficiency of a Geiger–Müller counter	120
	4.2 Determination of dead time of a Geiger–Müller counter	122
5	The statistics of radioactive decay	124
6	Determination of the energy of α-particles by measurement of their range in air	131
	6.1 Measurement of α-particle range using a Geiger–Müller counter or a solid-state detector	133
	6.2 Measurement of α-particle range using a pulse electroscope	134
	6.3 Absorption of α-particles in aluminium foils	135
	6.4 Determination of α-particle range using a spark counter	137
7	Determination of the range and energy of β-particles	140
8	Deflection of β-particles in a magnetic field	147
9	Deflection of α-particles in a magnetic field	149

Experiment	10	Scattering of α-particles by the nucleus of the atom	152
	11	Absorption of γ-radiation	159
	12	The inverse square law	162
	13	Observation of nuclear particle tracks in cloud chambers	164
	14	Gamma-ray scintillation spectrometry	167
	15	An intercomparison of detector efficiencies	172
	16	Self-absorption in β-particle sources	174
	17	Determination of the solubility of a slightly soluble salt	182
		17.1 Determination of the solubility of silver chloride in water and in sodium chloride solutions	183
		17.2 Determination of the solubility of strontium carbonate	185
	18	Separation of radon-220 from the thorium series and determination of its half-life using an ionization chamber	188
	19	Radiochemical separation of protoactinium-234 from the uranium series	190
	20	Radiochemical separation of lead-212 and bismuth-212 from the thorium series	195
	21	Radiochemical separation of praseodymium-144 from its parent, cerium-144	199
	22	Preparation of iodine-128 using a laboratory neutron source and the Szilard–Chalmers reaction to concentrate the activity	202
	23	The use of nuclear emulsions for the detection of α-particles	207
		23.1 A study of the relationship between uranium concentration and α-particle track density	208
		23.2 Determination of the half-life of uranium-238 using a nuclear emulsion for the detection of α-particles	209
		23.3 The observation of thorium stars in a nuclear emulsion	211
	24	Determination of the specific surface area of a lead sulphate precipitate using sulphur-35-labelled sulphate	213
	25	Determination of the partition coefficient for iodine between carbon tetrachloride and water	215
	26	Radiometric analysis	217
		26.1 Radiometric determination of magnesium	217
		26.2 Radiometric determination of lead	219
		26.3 Radiometric determination of copper in a copper/nickel solution	222

Experiment	27	Photosynthesis	224
		27.1 Fixation of carbon-14-labelled carbon dioxide by plants	225
		27.3 The technique of macroscopic autoradiography using plant material	227
		27.3 Fixation of inorganic carbon into simple sugars by plants	228
		27.4 Translocation of labelled material in plants	230
		27.5 The incorporation of inorganic carbon into algae	232
	28	Respiration	234
		28.1 Respiration in plants and animals	235
		28.2 The influence of temperature on the respiration of bacteria	238
	29	Mineral nutrition and ion movement across membranes	240
		29.1 The uptake and translocation of phosphorus-32-labelled phosphate in plants	241
		29.2 Determination of the rate of uptake of phosphate and sulphate ions by plants	244
		29.3 Investigation of the effects of inhibitors and competitors on the uptake of ions by excised roots	247
	30	The technique of microscopic autoradiography for studying the localization of radioisotopes in tissue	249
Appendix	1	Glossary of nuclear terms	254
	2	The thorium decay series	261
	3	The uranium decay series	262
	4	Radioactivity data for uranium and thorium	263
	5	Table of β,γ-emitting isotopes	264
	6	Lost counts corrections for Geiger–Müller counters	265
	7	Lost counts corrections for a paralysis time of 400 μs	266
	8	Decay corrections	267
	9	The electromagnetic spectrum	268
	10	Backscatter peak and Compton edge energies	269
	11	Some physical constants	271
Index			273

TABLES

1	Values of growth factor $(1 - e^{-\lambda t})$	24
2	Specific activities of some carrier-free radio-isotopes	25
3	Some radioisotope production reactions	27
4	Phosphors for scintillation counting	54
5	Some chemicals used in scintillation counting	55
6	Examples of the coding system on m-RNA	68
7	The linear energy transfer and quality factor for various radiations	76
8	Maximum permitted doses for various body organs	80
9	Some data on the internal deposition of isotopes	82
10	Classification of radionuclides according to toxicity	84
11	The shielding required for different types of radiation	87
12	Dose rates due to some Department of Education and Science approved sources	100
13	Maximum quantities of isotopes that may be used per experiment in schools	101
14	The probabilities of observing n "heads" in a throw of 16 coins	125
15	Variation of error in counting rate with number of counts recorded	127
16	The combination of errors in arithmetic operations	128
17	The limits of the quantity χ^2 for sets of counts with random errors	130
18	Energies and corresponding ranges in air of the α-particles from α-emitting sources	133
19	Variation of the velocity (v) and mass (m) of β-particles with their kinetic energy (E)	140
20	Range in aluminium of β-particles with kinetic energies from 0·15 MeV to 3·0 MeV	143
21	Impact parameters at various scattering angles for scattering of a 5 MeV α-particle by a gold nucleus	155
22	Some results for the α-particle scattering analogy	158
23	Gamma-ray sources for spectrometer calibration	170
24	Theoretical values of counting efficiency for sources with self-absorption	180
25	Part of the uranium decay series	190
26	The latter part of the thorium series	195
27	Neutron sources	203
28	Gamma-ray recoil energies for neutron capture and the Szilard–Chalmers effect in halides	204
29	Estimated exposure times for plant material labelled with carbon-14 for autoradiography	227

PLATES (at end of book)

1. An end-window Geiger–Müller counter in a lead castle, with a quench unit and a scaler incorporating an e.h.v. supply
2. An end-window Geiger–Müller counter in a Perspex castle with a scaler incorporating an e.h.v. supply
3. A Perspex castle with a Geiger–Müller counter for liquid samples and a chromatogram scanner
4. Solid-state detectors for α-particle counting
5. A laboratory bench prepared for work with open sources of radioisotopes
6. Examples of isotope packaging
7. A pulse electroscope with a radon source bottle
8. Apparatus for studying the deflection of β-particles in a magnetic field
9. An experimental analogy for studying the scattering of α-particles by heavy nuclei
10. The tracks of α-particles in a diffusion-type cloud chamber
11. Apparatus for the separation of lead-212 and bismuth-212 from the thorium series
12. Uranium-238 α-particle tracks in nuclear emulsion (see Experiment 21)
13. Alpha particle tracks in a nuclear emulsion due to the decay of radium-224 (see Experiment 21)
14. Macroscopic autoradiograph of a leaf labelled with carbon-14 by photosynthesis
15. Photomicrograph and autoradiograph of a section of the thyroid gland of a tadpole
16. The right lobe of the thyroid gland shown in Plate 15, under higher magnification

Suggestions for Course Work

The following experimental work is suggested as the basis on which courses for the disciplines of physics, chemistry and biology may be planned. It is improbable that time will permit all the work suggested to be carried out in any given course and a selection will have to be made. This selection we leave to the individual teacher concerned.

Physics course
Nos. 1–16, 18, 19, 20, 22, 23, 27.1, 27.2

Chemistry course
Nos. 1, 2, 4, 5, 7, 11, 12–17, 19 or 21, 22–27

Biology course
Nos. 1, 2, 4, 5, 7, 11–16, 19, 27–30

Radioactive materials
The Department of Education and Science approved closed (sealed) sources of α, β and γ-emitting radioisotopes are obtainable from the usual laboratory equipment suppliers. Closed sources of higher activity are supplied by the Radiochemical Centre Ltd.

Open sources of radioactive materials for use as solutions are generally obtainable from the Radiochemical Centre Ltd., Amersham, Bucks, and frequent reference to their catalogue has been made. Small amounts of radioactive materials as open sources may be obtained by means of the "tablet" scheme operated by J. A. Radley (Laboratories) Ltd., 220 Elgar Road, Reading, Berks.

Chapter 1
Nuclei and radioactivity

The atomic nature of matter, the nuclear structure of the atom and the existence of unstable or radioactive nuclei are ideas which are now so well established that they must be regarded as fundamental concepts of science. The atomic theory was not the idea of any one man, but the first clear statement of it is generally credited to John Dalton, who in 1803 stated that "Atoms are real discrete particles of matter which cannot be sub-divided by any known chemical process." This definition is still valid today.

Evidence that the atom might have an internal structure which could not be revealed by chemical processes accumulated during the latter part of the 19th century as a result of experimental work on the production of electrical discharges in gases and X-rays. Towards the end of the century, Henri Becquerel was investigating the phosphorescent properties of materials and found that uranium salts emitted penetrating radiations whether they had been exposed to light or not. He established that these radiations caused ionization of air and would discharge an electroscope in the same way as X-rays. These results were reported in 1896. The term *radioactivity* to describe the phenomenon was later introduced by Pierre and Marie Curie. The Curies found that uranium minerals showed greater radioactivity than chemically separated uranium salts, and undertook a search for other radioactive substances in uranium minerals. This led to the discovery of polonium and radium in 1898.

By this time, Rutherford had shown that the radiations emitted by uranium were of two kinds, the easily absorbed α-particles, which we now know are charged helium atoms, and the more penetrating β-particles which are high-energy electrons. Later a third component, highly penetrating electromagnetic radiation, or γ-rays, was discovered. The nature of all these types of radiation was established by experiments to determine the charge and mass of the particles. The identity of α-particles with helium atoms was demonstrated spectroscopically by Ramsey and Royds in 1909.

The next major advance came in 1911, when Rutherford published his nuclear theory of the atom. Rutherford, with his co-workers Geiger and Marsden, had been investigating the scattering of α-particles by thin foils of gold, copper and aluminium. The foils were about 10^{-5} cm thick, which corresponds to many atomic diameters (about 100 diameters in the case of gold). It was observed that, although most of the α-particles passed through the foil without deflection, some were deflected from their initial direction by small angles and a very few (about 1 in 8000) were deflected through large angles. Some of these α-particles were even scattered back along their incident path. To explain these observations, Rutherford postulated that the atom must include a very small nucleus, about $\frac{1}{10000}$ of the diameter of the atom, in which nearly all of the mass of the atom resides. This nucleus is surrounded by a cloud of electrons which occupy most of the space in the atomic volume. The nucleus carries a positive charge, equal to Z times the charge on the electron, where Z is known as the *atomic number* of the atom. So that the atom will be electrically neutral, there are Z electrons in the surrounding electron cloud. The mass of the electron was by this time known to be only $\frac{1}{1840}$ of the mass of the hydrogen atom, so it was inconceivable that the heavy α-particle would be deflected from its incident path on collision with such a very light particle.

The deflection of an α-particle could occur only if it came into the region of high electric field close to a positively charged nucleus. The resulting electrostatic repulsion would cause the relatively light α-particle to be deflected without disturbing the position of the nucleus. Such an interaction would be a rare event, since the chances of an α-particle colliding with the very small nucleus are small compared to the chances of an interaction in the relatively large electron cloud surrounding the nucleus. On interaction with electrons the α-particle loses some of its kinetic energy but it is not deflected from a straight-line path. Rutherford, by studying the angular distribution of the α-particles scattered by the nuclei of heavy atoms, was able to deduce the actual sizes of nuclei. He found that most nuclei have a radius of the order of 10^{-13} cm; the method by which this may be derived from α-particle scattering measurements is discussed in Experiment 10.

The relative weights of the atoms of the various elements had been determined by making use of the laws of chemical combination, early in the 19th century. These *atomic weights* were at first determined relative to the hydrogen atom. Later in the century, atomic weights were determined relative to that of oxygen, which was arbitrarily taken as being 16·0000 units. Evidence of periodicity in the chemical properties of the elements enabled Mendeleef to publish his Periodic Table of the Elements in 1869. In this table, the elements were arranged, with a few exceptions, in the order of their atomic weights. This order was later confirmed by Moseley, in 1914, who established the order from the sequential nature of the X-ray

Nuclei and Radioactivity

spectra which he observed for the various elements. The position of an element in the periodic table enabled a serial number, the atomic number, to be assigned to each element. As a result of Rutherford's work on α-particle scattering, and Moseley's work on X-ray spectra, it became clear that the nuclear charge of the atom was equal to its atomic number.

The nuclear nature of the atom having been established by Rutherford, the structure of the atomic nucleus itself posed the next problem. It was logical to assume, as a starting point for any hypothesis concerning nuclear structure, that the nucleus of the lightest atom, hydrogen, called a proton, was a simple nuclear building block. This, however, could not be the only nuclear component since, for example, a nucleus consisting of 12 protons would have the correct mass for a carbon atom but would carry 12 positive charges, double the charge on the carbon nucleus. Therefore in 1920, Rutherford postulated the existence of another nuclear building block, the neutron, which could be regarded as a close combination of a proton and an electron. The neutron would therefore have zero charge and a mass slightly greater than that of the proton. The structure of the carbon atom would thus be:

Extra-nuclear particles
   Electrons   6
   Charge $6 \times e^-$ coulomb*
   Mass $6 \times \frac{1}{1840}$ u†

Nucleus
   Protons   6
   Charge $6 \times e^+$ coulomb
   Neutrons   6
   Mass   12 u

* $e = \pm 1 \cdot 602 \times 10^{-19}$ C
† see page 4.

The conventional way to represent an atom of element X is $^A_Z X$, the superscript A being the *mass number* of the atom (the number of nuclear components or nucleons in its nucleus), and the subscript Z being the atomic number of the atom (the number of protons in the nucleus). For the atom of carbon, the symbol would therefore be $^{12}_6 C$.

Rutherford's concept of the neutron also led to an explanation of his and Soddy's conclusion that amongst the natural radioactive elements there exist atoms of an identical chemical nature but with different masses. These are called *isotopes*. For example, $^{210}_{82}Pb$ occurs in the decay of uranium, whilst $^{212}_{82}Pb$ is found in the decay products of thorium. These two nuclei each contain 82 protons but 128 and 130 neutrons respectively. The atoms are chemically identical since each contains 82 electrons in the extra-nuclear electron cloud, and it is these electrons which determine their chemical properties.

It was not until 1932, that Chadwick found experimental evidence for the existence of the neutron. He observed a very penetrating type of radiation which was emitted when boron was bombarded with α-particles, and succeeded in showing that this radiation consisted of neutrons produced by the reaction

$$^{11}_{5}B + ^{4}_{2}He \rightarrow ^{14}_{7}N + ^{1}_{0}n \qquad (1.1)$$

It is now known that such a free neutron, which is not associated with a nucleus, will decay spontaneously to form a proton and an electron, thus:

$$^{1}_{0}n \rightarrow ^{1}_{1}p + e^{-} \qquad (1.2)$$

Atomic weights have been determined for more than a century on the basis of the oxygen scale. For chemical purposes, such as the calculation of the masses taking part in a chemical reaction, this scale was perfectly satisfactory. However, in the case of nuclear reactions, in which a very small mass change can lead to very large energy release for a reaction, it was necessary to define a much more precise mass scale. The atomic weight had as its basis the average weight of an oxygen atom in the natural mixture of isotopes. (Natural oxygen contains 99·76% of $^{16}_{8}O$, 0·04% of $^{17}_{8}O$ and 0·2% of $^{18}_{8}O$). The standard selected for an *atomic mass scale* was the mass of the oxygen-16 isotope, which was taken as 16·000 000 units. To avoid possible confusion in the use of different mass scales for atomic weights and nuclear or isotopic masses, a new common basis for both scales was introduced by international agreement in 1959. The basis of this scale is the mass of the carbon-12 isotope, which is taken as equal to 12·000 000 units. The unit of both the atomic weight scale and the atomic mass scale is then defined as $\frac{1}{12}$th of the mass of the carbon-12 isotope. This unit, is known as the *unified atomic mass unit* (abbreviation, u, formerly a.m.u.).

NUCLEAR STABILITY AND NUCLEAR FORCES

The spontaneous decay of a neutron to a proton and an electron also takes place within certain nuclei, namely those radioactive nuclei which decay by β-particle emission. Before discussing such radioactive decay processes in more detail, it is useful to consider some general aspects of nuclear stability.

Nuclei and Radioactivity

The forces which hold the components of a nucleus together must operate over extremely small distances, 10^{-13} cm or less. It is now known that these forces arise because of the exchange of particles called *π-mesons*, between the constituent particles, or *nucleons*, in a nucleus. These mesons may be neutral or may carry a positive or a negative charge. Exchange of charged mesons may take place between protons and neutrons. Transfer of a neutral meson may take place from one proton to another, or from one neutron to another. Such exchange processes result in quantum-mechanical exchange forces which will bind the nucleons together. The nucleons in a nucleus may be regarded as existing in a meson gas, and these exchange processes are continually taking place resulting in the generation of the forces which hold the nucleons together.

The binding energy of a nucleus

Another concept which enables the magnitude of nuclear forces to be estimated is the *binding energy* of the nucleus. A nucleus is less massive than the sum of the masses of its component nucleons. This arises because, as separated nucleons are brought together to form a nucleus, energy is expended in forming the bonds between the nucleons, and this results in a loss of mass equivalent to the energy involved. Conversely, the binding energy of a nucleus is the energy required to separate its constituent nucleons and it may be calculated from the masses of the proton, the neutron and the nucleus concerned, making use of Einstein's relationship between mass and energy:

$$E = mc^2 \qquad (1.3)$$

This relationship gives the energy equivalent of mass m, where c is the velocity of light.

The *average binding energy per nucleon* is the nuclear binding energy divided by the number of nucleons. It may be regarded as a measure of the nuclear forces which bind the nucleons together for that particular nucleus. The variation of the average binding energy per nucleon with the nuclear size or mass number is shown in Fig. 1.1

For dealing with nuclear changes it is useful to evaluate this relationship for the case where m is expressed in atomic mass units and energy is expressed in mega-electronvolts. The actual mass equal to 1 u is $\frac{1}{12}$ of the mass of a carbon-12 atom:

$$1\,\text{u} = \frac{1}{12} \times \frac{12}{6 \cdot 022\,169 \times 10^{26}} = 1 \cdot 660\,53 \times 10^{-27}\,\text{kg}$$

since one kilomole of carbon contains $6 \cdot 022\,169 \times 10^{26}$ atoms (Avogadro number.) The *electronvolt* (eV) is defined as the energy acquired by an electron when it is accelerated in an electric field through a potential difference of 1 V. This is a very small amount of energy, equivalent to $1 \cdot 6 \times 10^{-19}$ J. It would require $2 \cdot 6 \times 10^{13}$ eV to raise the temperature of 1 cm³ of water by 1°C. Expressing m in atomic mass units and E in mega-electronvolts (10^6 eV) the Einstein

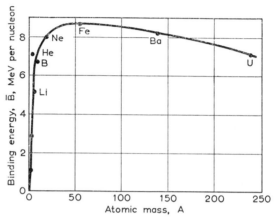

Fig 1.1 Variation of the average binding energy per nucleon with mass number

equation shows that the energy equivalent of 1 u is 931 MeV. This is the quantity of energy released for each atomic mass unit that is converted to energy in nuclear reactions such as fusion and fission or in nuclear decay processes. In more familiar terms, the energy necessary to boil 1000 cm³ of water would be obtained by conversion of about $3 \cdot 7 \times 10^{-13}$ g of matter to energy.

Calculation of the binding energy of a nucleus

Suppose that the nucleus $^A_Z X$ is separated into its constituent nucleons in the imaginary reaction

$$^A_Z X + Q = Z ^1_1 p + (A-Z) ^1_0 n \tag{1.4}$$

where Q is the total *binding energy* of the nucleus. This is the energy that must be supplied in order to overcome the nuclear forces in the process. Substituting the energy equivalents of the masses (eqn. (1.3)) involved in this equation gives

$$M_A c^2 + Q = Z m_p c^2 + (A-Z) m_n c^2 \tag{1.5}$$

where M_A is the mass of the nucleus and m_p and m_n are the masses of the proton and neutron respectively. The average energy, $\bar{B}$, required to

Nuclei and Radioactivity

remove a nucleon from the nucleus is then given by

$$\bar{B} = Q/A = \{Z \cdot m_p c^2 + (A - Z)m_n c^2 - M_A c^2\}/A \tag{1.6}$$

Since nuclear masses can be accurately determined and the masses of the proton and neutron have been established from nuclear decay data, $\bar{B}$ may be evaluated for any nucleus. The variation of the average binding energy per nucleon, $\bar{B}$, with atomic mass, A, is shown in Fig. 1.1.

ENERGY RELEASE IN NUCLEAR REACTIONS

It may be noted that the nuclear binding forces are strongest around mass numbers 50 to 70 and that they diminish rapidly for mass numbers less than 50 and decrease more slowly for mass numbers greater than 70. The two ways in which considerable amounts of energy may be released by nuclear reactions are now readily understood. The first, the fusion of light nuclei, may be represented by the reaction

$$^2_1\text{H} + ^3_1\text{H} \rightarrow ^4_2\text{He} + ^1_0\text{n} \tag{1.7}$$

in which the nucleus of the stable isotope of hydrogen, deuterium, reacts with the nucleus of its radioactive isotope, tritium. The energy required to separate the nuclei involved in this reaction into their constituent nucleons may be calculated from the known binding energies (B.E.) for masses 2, 3 and 4, as follows:

$$
\begin{aligned}
\text{Total B.E. of reactants} &= 2 \times 1\cdot09 \text{ (B.E. mass 2)} \\
&\quad + 3 \times 2\cdot53 \text{ (B.E. mass 3)} \\
&= 9\cdot77 \text{ MeV} \\
\text{Total B.E. of products} &= 4 \times 7\cdot03 \text{ (B.E. mass 4)} \\
&= 28\cdot12 \text{ MeV} \\
\text{Change in B.E.} &= 18\cdot35 \text{ MeV}
\end{aligned}
$$

Since the total binding energy of the products is greater than that of the reactants, kinetic energy equal to the difference, 18·35 MeV, will be released for each such nuclear fusion that takes place. If all the deuterium present in 1000 cm³ of heavy water, $^2\text{H}_2\text{O}$, underwent this reaction it would produce sufficient energy to boil 5×10^8 kg of water. Heavy water can be separated from natural water at a cost of about £20 per 1000 cm³, but the technological problems involved in harnessing this source of energy have not yet been solved.

The second means of deriving energy from nuclear reactions is to make use of the fission reaction which is the basis of nuclear reactor operation. In this reaction, heavy nuclei are split into two lighter

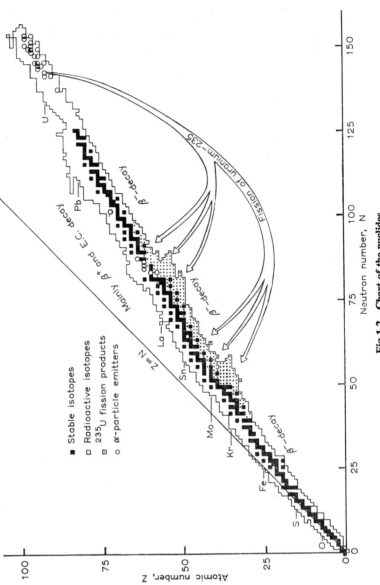

Fig 1.2 Chart of the nuclides

Nuclei and Radioactivity

fragments as, for example, in the neutron-induced fission of uranium-235. The following reaction shows one example of the many ways in which the uranium atom may be divided in the fission process.

$$^{235}_{92}U + ^{1}_{0}n = ^{144}_{56}Ba + ^{90}_{36}Kr + 2^{1}_{0}n \qquad (1.8)$$

The change in the nuclear binding energy when this reaction takes place may be calculated as follows:

Total B.E. of reactants = 235 × 7·6 (B.E. mass 235)

= 1786 MeV

Total B.E. of products = 144 × 8·26 (B.E. mass 144)

+ 90 × 8·6 (B.E. mass 90)

= 1962 MeV

Change in B.E. = 176 MeV

This amount of energy is released mainly as kinetic energy for each atom of uranium-235 that undergoes fission. The energy released by fission of all the uranium-235 atoms in 1 kg of natural uranium would be sufficient to boil about $1·56 \times 10^8$ kg of water and the cost of this quantity of uranium would be about £33. The cost of nuclear power is, of course, very much more than this since the uranium cost is not the only one involved. At present, to boil this amount of water by electricity generated in a nuclear power station would cost about £4500.

THE NUCLIDES CHART

All the known nuclei, both stable and radioactive, may be represented graphically by plotting their atomic number, Z, against their neutron number, N, to the same scale as shown in Fig. 1.2. The stable nuclei are represented by black squares, and it may be noted that these lie on a "stability line", which for atomic numbers up to about 20 has a 45° slope indicating that Z equals N. Above an atomic number of 20, the slope of the stability line decreases as N becomes greater than Z for elements of higher atomic number. Unstable nuclei lie below the stability line if their neutron number is too great for stability, and those above the stability line have a neutron number too small for stability. On such a chart, the isotopes of an element lie along the horizontal lines, and *isobars* (nuclei of the same mass) lie along diagonal lines.

MODES OF RADIOACTIVE DECAY

The nuclear changes which take place on radioactive decay may be indicated on the nuclide chart, as shown in Fig. 1.3.

α-decay

Decay by α-particle emission occurs mainly amongst the nuclei of atomic number greater than that of lead, and involves the emission of a helium nucleus from the nucleus of the parent atom; for example,

$$^{226}_{88}Ra \rightarrow {}^{222}_{86}Rn + {}^{4}_{2}He^{++} \qquad (1.9)$$

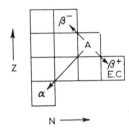

Fig 1.3 Changes in neutron number and proton number for various decay modes

This is represented in Fig. 1.3 by a 2-step diagonal movement towards lower Z and lower N, two protons and two neutrons having been lost from the parent nucleus.

β-decay

The β-particle is an energetic electron released from the nucleus in the process of β⁻-decay. It may be regarded as having been formed by the change of a neutron to a proton within the parent nucleus, with the emission of a β⁻-particle and another particle, an *antineutrino*; thus

$$^{1}_{0}n \rightarrow {}^{1}_{1}p + e^{-} + \bar{\nu} \quad \text{(antineutrino)} \qquad (1.10)$$

This will cause the formation of a new nucleus of atomic number one unit greater; for example

$$^{32}_{16}P \rightarrow {}^{32}_{17}S + \beta^{-} + \bar{\nu} \qquad (1.11)$$

Such a decay is represented in Fig. 1.3 by a diagonal step along the isobar towards greater Z, the parent nucleus having apparently lost a neutron and gained a proton. The antineutrino and the *neutrino*, which will be mentioned later, are fundamental particles which have zero electric charge and an extremely small mass, the magnitude of which depends upon their kinetic energy. They can only be detected by the interactions which they make with the nuclei of certain atoms. These interactions are extremely rare; neutrinos from the sun, for example, can penetrate deep into the earth and even pass through the earth without making any such interactions.

β^+-decay and K-electron capture

The *positron* is a positively charged electron which is emitted in some nuclear decay processes and can also be formed in the absorption of high-energy γ-rays. The decay of a radioisotope by positron (β^+) emission may be regarded as a change of a proton within the parent nucleus to a neutron, a positron and a neutrino; thus

$$_1^1p \rightarrow {}_0^1n + e^+ + \nu \tag{1.12}$$

This will cause the formation of a new nucleus of atomic number one unit less; for example,

$$_{11}^{22}Na \rightarrow {}_{10}^{22}Ne + \beta^+ + \nu \tag{1.13}$$

Such a decay is represented in Fig. 1.3 by a diagonal step along the isobar towards smaller Z, the parent nucleus having apparently lost a proton and gained a neutron.

The decay of a nuclide by *K-electron capture* (E.C.) involves the capture by the nucleus of an electron from the innermost (K) shell of extra-nuclear electrons, and the change in the nucleus is

$$_1^1p + e^- \rightarrow {}_0^1n + \nu \tag{1.14}$$

This also leads to the formation of a new nucleus whose atomic number is one unit less; for example,

$$_{24}^{51}Cr + e^- \rightarrow {}_{23}^{51}V + \nu \tag{1.15}$$

and is represented in Fig. 1.3 by a movement similar to that for positron decay.

It may be noted that those nuclides which undergo β^- decay generally lie below the stability line of Fig. (1.2). They are in fact nuclei which have an excess of neutrons; their neutron-to-proton ratio is too high for stability. The nuclides which decay by β^+ emission and electron capture, on the other hand, are neutron deficient; their neutron-to-proton ratio is too low for stability. In all cases the decay of the nuclide takes it closer to the stability line.

THE ENERGY RELEASED IN DECAY BY α-PARTICLE EMISSION

The α-particle decay process may be generalized as

$$^A_Z X \rightarrow ^{A-4}_{Z-2} Y + ^4_2 He^{++} + Q \quad (1.16)$$

where Q is the disintegration energy. The process will occur spontaneously only if Q is positive; that is, if energy equal to Q is released as a consequence of the decay.

Calculation of the disintegration energy for α-emission

It is possible to calculate the value of Q from the mass change, Δm, in the decay process using the Einstein relationship between mass and energy as follows:

$$Q = \Delta m c^2 = (m_Z - m_{Z-2} - m_\alpha)c^2 \quad (1.17)$$

The reaction involves the nuclei of the atoms only; the masses m_Z, m_{Z-2} and m_α are the nuclear masses of the atoms X, Y and the α-particle respectively. These masses may be converted to the more usual atomic masses by adding the appropriate number of electron masses to each, for example, the atomic mass of $^A_Z X$ is the sum of the mass of its nucleus and the extra-nuclear electrons, and is given by

$$M_Z = m_Z + Z m_e$$

where m_e is the mass of the electron. The energy released in the process of α-decay therefore becomes

$$Q = m_Z c^2 + Z m_e c^2 - (m_{Z-2} c^2 + (Z-2) m_e c^2 + m_\alpha c^2 + 2 m_e c^2) \quad (1.18)$$

$$= M_Z c^2 - M_{Z-2} c^2 - M_{He} c^2 \quad (1.19)$$

It may be noted that there has been no net addition of electron mass to this equation.

Decay diagrams for α-emission

The energy changes occurring on α-decay may be represented diagrammatically by plotting the energy equivalents of the atomic masses of the parent and daughter nuclides on a vertical scale. The line representing the daughter state is conventionally shown to the left of the parent state to indicate the decrease in atomic number which occurs on α-emission, as in Fig. 1.4. The energy difference between the parent and daughter states is the disintegration energy, Q, for the decay process. So that the law of conservation of momentum can be preserved, the disintegration energy must be divided between the α-particle emitted and the daughter, or recoil, nucleus.

The kinetic energy of the α-particle, E_α, is related to the disintegration energy by the equation

$$E_\alpha = \frac{m_{Z-2}}{m_\alpha + m_{Z-2}} Q \quad (1.20)$$

Nuclei and Radioactivity

The daughter may often be left in various alternative energy states following α-decay. A radionuclide which decays by α-emission may emit α-particles of more than one energy, depending upon the number of energy states in which the daughter may be formed. The decay of radium-226, for example, is illustrated in Fig. 1.4, where it may be noted that the daughter, radon-222, is left in an excited

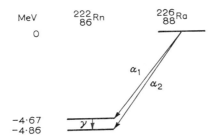

Fig 1.4   Decay of radium-226

Energy of α_1	4·59 MeV (5·7%)
Energy of α_2	4·78 MeV (94·3%)
Energy of γ-ray	0·19 MeV
Half-life	1620 y

state in 5·7% of the disintegrations of radium. In this diagram the energy of each state has been plotted taking the energy equivalent of the mass of the parent, radium-226, as an arbitrary zero. For the disintegration to the ground state, the disintegration energy is 4·86 MeV and the energy of the α-particle is 4·78 MeV. The energy difference, 0·08 MeV, is the kinetic energy of the radon recoil atom. The lifetime of an excited state of a nucleus is extremely short, less than 10^{-14} s, and immediately following decay to the excited state of radon-222, a γ-ray photon of energy 0·19 MeV is emitted, leaving the nucleus in its ground (or lowest) energy state. The diagram summarizes the decay of radium-226 completely, but it must be remembered that the daughter, radon-222, is also radioactive.

THE ENERGY RELEASED IN DECAY BY β^--PARTICLE EMISSION

Nuclear decay by emission of a β^--particle may be generalized as

$$^{A}_{Z}X \rightarrow ^{A}_{Z+1}Y + \beta^- + \bar{\nu} + Q \tag{1.21}$$

where Q is the disintegration energy for the process. Since the mass of the recoil nucleus Y is very much greater than that of the electron, the recoil energy of Y is negligible and the entire disintegration energy Q appears as kinetic energy of the β^--particle and the antineutrino. This energy may be partitioned between the β^--particle and the antineutrino in any manner, so that a plot of the number of β-particles of energy E against E gives a β-particle spectrum of the kind shown in Fig. 1.5. A very few β⁻-particles have very little kinetic energy so that the associated antineutrino must have almost the total disintegration energy Q; likewise a few β⁻-particles have an energy equal to the disintegration energy, and the associated antineutrino

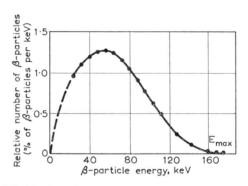

Fig 1.5 The β-particle spectrum of sulphur-35 measured with a β-particle spectrometer

in these cases must have zero energy. Between these two extremes there is a continuous spectrum of β⁻-particle energies. The existence of this distribution of β⁻-particle energies was one of the reasons for postulating the existence of the antineutrino.

Calculation of the disintegration energy for β⁻-particle emission

The value of Q may be calculated from the equation

$$Q = m_Z c^2 - m_{Z+1} c^2 - m_e c^2 \tag{1.22}$$

which is eqn. (1.21) expressed in nuclear mass terms.

Converting nuclear masses to atomic masses yields

$$Q = M_Z c^2 - M_{Z+1} c^2 \tag{1.23}$$

so that the disintegration energy may be calculated from the masses of the parent and daughter isotopes.

Decay diagrams for β⁻ emission

The disintegration energy for β-particle decay is equal to the difference in the energy equivalents of the atomic masses of the parent and daughter nuclides. The decay of praseodymium-144 is illustrated

Nuclei and Radioactivity

in Fig. 1.6. As with α-decay the energy equivalent of the mass of the parent nuclide has been taken as zero. The daughter, neodymium-144, is sometimes formed in an excited state and in such cases the excited state decays immediately to the ground state with the emission of a γ-ray photon. Conventionally, the daughter state is shown to the right of the parent to indicate the increase in atomic

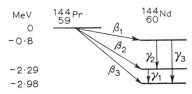

Fig 1.6   Decay of praseodymium-144

Energy of β_1	0·8 MeV	(1%)
Energy of β_2	2·29 MeV	(1%)
Energy of β_3	2·98 MeV	(98%)
Energy of γ_1	0·69 MeV	(1·6%)
Energy of γ_2	1·49 MeV	(0·3%)
Energy of γ_3	2·18 MeV	(0·8%)
Half-life	17 min	

number which occurs in the β^--decay process. Since the mass of the β^--particle is negligible compared to that of the daughter nucleus, the recoil energy of the daughter is negligible and the maximum energy of the β^--particles emitted is equal to the disintegration energy for the decay.

THE ENERGY RELEASED IN DECAY BY β^+-PARTICLE (POSITRON) EMISSION

Nuclear decay by positron emission may be generalized as

$$^A_Z X \rightarrow ^A_{Z-1} Y + \beta^+ + \nu + Q \tag{1.24}$$

where Q is the disintegration energy, which is released as the kinetic energy of the positron and the neutrino, since the mass of the positron is negligible compared with that of the recoil nucleus. The energy is partitioned between the positron and the neutrino so that the positrons emitted will have an energy spectrum similar to that observed for β^--particles.

Calculation of the Disintegration energy for positron emission

The value of Q may be calculated from the equation

$$Q = m_Z c^2 - m_{Z-1} c^2 - m_e c^2 \tag{1.25}$$

which is eqn. (1.24) in nuclear mass terms.

On conversion of nuclear masses to atomic masses this equation becomes

$$Q = M_Z c^2 - M_{Z-1} c^2 - 2m_e c^2 \tag{1.26}$$

The disintegration energy for positron decay is thus given by the difference between the energy equivalents of the masses of the parent and daughter nuclides, less 1·02 MeV, the energy equivalent of two electron masses.

Decay diagrams for β^+-emission

The disintegration energy for positron emission is equal to the energy equivalent of the mass difference between the parent and

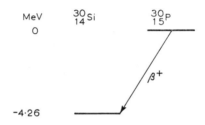

Fig 1.7 Decay of phosphorus-30

Positron energy   4·26 − 1·02 = 3·24 MeV
Half-life   2·6 min

daughter nuclides, less the energy equivalent of two electron masses, namely 1·02 MeV. The decay of the positron emitter, phosphorus-30, is represented in Fig. 1.7. The daughter state is drawn to the legs of the parent state to indicate the decrease in atomic number which occurs on positron emission.

THE ENERGY RELEASED IN DECAY BY K-ELECTRON CAPTURE

The process by which an unstable nucleus decays by capture of a K-shell electron from the cloud of extra-nuclear electrons may be represented as

$$_Z^A X + e^- \rightarrow {}_{Z-1}^A Y + \nu + Q \tag{1.27}$$

In this decay the disintegration energy, Q, will appear as kinetic energy of the neutrino or as γ-ray energy if the daughter is left in an excited state. The neutrino can only be detected as a result of its nuclear interactions, which are very rare and will not therefore be detected by the usual methods used for radioactivity measurements. The daughter nucleus $_{Z-1}^A Y$ will initially be formed with a vacancy in

Nuclei and Radioactivity

its K-shell of electrons, and as the electrons rearrange themselves to fill this vacancy, X-rays characteristic of the element Y will be emitted. Nuclei which decay by electron capture may often be detected by either the γ-radiation emitted if the daughter is formed in an excited energy state, or by the X-radiation which is emitted following electron capture.

Calculation of the disintegration energies for K-electron capture

The disintegration energy for electron capture in nuclear mass terms is given by the equation

$$Q = m_Z c^2 - m_{Z-1} c^2 + m_e c^2 \tag{1.28}$$

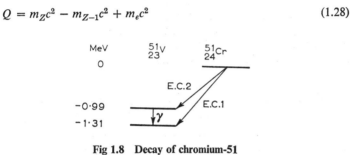

Fig 1.8 Decay of chromium-51

E.C.1 Neutrino energy	1·31 MeV	(90%)
E.C.2 Neutrino energy	0·99 MeV	(10%)
E_γ	0·32 MeV	
Half-life	27·8 days	

On conversion of nuclear masses to atomic masses this equation becomes

$$Q = M_Z c^2 - M_{Z-1} c^2 \tag{1.29}$$

The disintegration energy is therefore equal to the energy equivalent of the mass difference between the parent and daughter nuclides.

Decay diagrams for K-electron capture

The disintegration energy for the decay is given by the difference in the energy equivalents of the masses of the parent and daughter nuclides.

A decay diagram for the electron-capture isotope, chromium-51, is shown in Fig. 1.8. In 10% of the chromium-51 electron-capture events the daughter, vanadium-51, is left in an excited state. The immediate decay of this state to the ground state leads to the emission of 0·32 MeV γ-ray photons. The other 90% of the capture events lead directly to the ground state of vanadium, and the disintegration energy appears as kinetic energy of the neutrino. All the electron-capture events are followed by the filling of the electron vacancy in the K-shell of vanadium, and therefore the characteristic

0·005 MeV vanadium X-rays are also observed. In transitions to the ground state, the emission of these X-rays will be the only observable evidence that decay has occurred.

GAMMA RAY EMISSION

As we have seen in the above discussion, the emission of γ-radiation generally follows some other nuclear-decay process. It is not itself a primary nuclear-decay process, but is emitted when a transition occurs between an excited state of a nucleus and a lower nuclear-energy state. These excited states are associated with the daughter nucleus formed after α-particle or β-particle decay processes. Generally, the lifetimes of excited states are extremely short so that γ-ray emission immediately follows the decay process which led to the production of the excited states.

THE DISINTEGRATION RATE OF RADIOISOTOPES

The rate at which radioactive nuclei disintegrate is dependent upon only two factors: the identity of the radionuclide involved, and the number of nuclei which are present. This decay rate, dN/dt, is given by the equation

$$\frac{dN}{dt} = -\lambda N \tag{1.30}$$

where N is the number of radioactive nuclei present, and λ is the radioactive decay constant.

The value of the radioactive decay constant is characteristic of each particular radioisotope. A relationship between N_0, the number of atoms of the radioisotope which are present initially, and N_t, the number of atoms present after time t, is obtained on integration of eqn. (1.30). This is discussed more fully in Experiment 1, but one equation expressing this relationship is

$$\log_e N_0/N_t = \lambda t \tag{1.31}$$

The *half-life*, $t_{1/2}$, of a radioisotope is a most useful quantity for describing its decay rate. This is defined as the time which must elapse to reduce by half the number of atoms initially present. Substituting $N_t = N_0/2$ in eqn. (1.31) yields the following relationship between the half-life, $t_{1/2}$ and λ:

$$\log_e 2 = \lambda t_{1/2}$$

or

$$\frac{0 \cdot 693}{\lambda} = t_{1/2} \tag{1.32}$$

Nuclei and Radioactivity

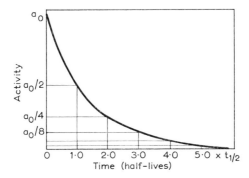

Fig 1.9 Decay of a radioisotope

The half-lives of radioisotopes vary from very short, about 10^{-4} s in the case of polonium-214, for example, to extremely long, $1 \cdot 3 \times 10^9$ years for potassium-40. The number of atoms of a radioisotope decreases by half after the elapse of each half-life in the way shown in Fig. 1.9.

Chapter 2

Radioisotopes and labelled compounds

A very large number of nuclei, both radioactive and stable, have now been identified. The only radioactive nuclei that occur naturally to-day are those with very long half-lives. These include potassium-40 and isotopes of the elements from lead to uranium in the periodic table. Most of the natural α-particle emitters are to be found amongst these elements of high atomic number.

All the naturally occurring elements, both stable and radioactive, have been formed by the nuclear building processes that are known to take place in stars. After their formation in the interior of stars these elements subsequently became part of the interstellar dust and were eventually incorporated in the primordial planets. Only the long-lived radioactive isotopes are to be found in terrestrial rocks to-day, and it is obvious that the half-lives of these radioisotopes must be comparable in magnitude with the age of the earth. The radioactive heavy nuclei of shorter half-lives which are found in nature are decay products of uranium and thorium. Those radioactive nuclei which are no longer found in nature must be prepared by synthetic methods and are known as *artificial* radioisotopes. They are artificial only in that they are not found in nature to-day but must be man-made. The methods by which these man-made isotopes can be produced are discussed below.

NUCLEAR REACTIONS

The reactions which produce nuclei in stars are of two kinds, namely *fusion* reactions and *neutron-capture* reactions. In a fusion reaction, two nuclei collide and fuse to form a heavier nucleus. The colliding nuclei require considerable kinetic energy to overcome the very strong Coulomb repulsion forces due to their nuclear charges. It is only at temperatures of many millions of degrees Celsius that nuclei have sufficient kinetic energy for fusion reactions to take place. On the earth, it has as yet only been possible to cause the fusion of nuclei of low atomic number, in the high temperatures produced by nuclear explosions. The thermonuclear explosion which is thus initiated derives its energy from reactions like

$$^2_1H + ^3_1H \rightarrow ^4_2He + ^1_0n + \text{energy} \tag{2.1}$$

Reactions between heavy and light nuclei can be induced by accelerating the light nucleus, in a machine such as a cyclotron, until it has gained sufficient kinetic energy to overcome the repulsion forces as it approaches the heavy nucleus. The accelerated nucleus, which may be a proton, a deuteron ($^2H^+$) or sometimes an α-particle, is then allowed to collide with a target containing the other reactant nuclei. Magnesium, for example, which is bombarded with fast deuterons is transmuted to sodium according to the equation

$$^{24}_{12}Mg + ^2_1H^+ \rightarrow ^{22}_{11}Na + ^4_2He^{++} \tag{2.2}$$

or, in abbreviated form, $^{24}_{12}Mg\,(d, \alpha)\,^{22}_{11}Na$.

The sodium formed may subsequently be chemically separated from the magnesium target, and we may note that, since the only isotope of sodium formed in the target is sodium-22, every atom of the sodium thus separated is a radioactive one.

The second kind of nuclear reaction which takes part in nucleus building in stars is the neutron-capture reaction, and this can readily be induced on earth. The source of the neutrons for such reactions in stars is uncertain, but in stellar explosions like supernova it is thought that large numbers of neutrons are produced by γ-ray-induced neutron emission from nuclei. The most useful source of neutrons available to man is the nuclear reactor, in which the nuclei of uranium atoms are split into two smaller nuclei accompanied by the emission of a few neutrons (see eqn. (1.8)). Stable elements placed in a reactor are subjected to bombardment by these neutrons and undergo reactions such as

$$^{23}_{11}Na + ^1_0n \rightarrow ^{24}_{11}Na + \gamma \tag{2.3}$$

which may be abbreviated to

$^{23}_{11}Na\,(n, \gamma)\,^{24}_{11}Na$

This is known as a *neutron-capture reaction*, and the γ-ray which is emitted, called the *capture gamma*, is released at the instant of neutron capture by the target nucleus. We may note that in this kind of reaction the product isotope is chemically identical with the target. Thus, in the above example, it is not possible to separate the sodium-24 formed from the sodium target. Since in all cases only a very small fraction of the target atoms undergo neutron capture, a radioisotope produced in this way will contain a preponderance of inactive target atoms.

The nuclear reactor also provides man with another means of obtaining radioisotopes. As mentioned above, the nuclear reactor depends for its operation on the splitting, or fissioning, of the uranium atom. The fission reaction is induced by the interaction of neutrons with the uranium-235 nucleus as, for example, in the reaction

$$^{235}_{92}U + ^{1}_{0}n \rightarrow ^{236}_{92}U \rightarrow ^{138}_{53}I + ^{95}_{39}Y + 3^{1}_{0}n \qquad (2.4)$$

Only atoms of the rare uranium isotope, uranium-235, are fissioned, and there are a large number of radioisotopes formed as fission products, the most abundant of which have masses near 90 and 140 mass units. Strontium-90, iodine-131 and caesium-137 are examples of fission products in these two mass groups. The fission products can be chemically separated from the used uranium fuel elements of nuclear reactors. Since there are no inactive isotopes of these elements in the nuclear fuel, radioisotopes prepared in this way will not contain stable isotopes of the same element.

The rate of nuclear reactions

The rate, X, at which radioactive atoms are produced in nuclear reactions is given by

$$X = N_{tg}\sigma\phi \qquad (2.5)$$

where N_{tg} = Number of target nuclei in sample
σ = Cross-section for the reaction, in barns (see below)
ϕ = Particle flux to which the sample is subjected, in particles/cm²-second

For an element of atomic weight A, the number of target nuclei is given by

$$N_{tg} = \frac{w}{A}f \times 6\cdot02 \times 10^{23} \text{ atoms} \qquad (2.6)$$

where w = Mass of target element in grammes
f = Fractional isotopic abundance of the isotope involved in the reaction

The cross-section for the reaction (σ) is the probability that the nuclear reaction will take place when the bombarding particle approaches the target nucleus. The unit used to measure these reaction cross-sections is the *barn*, equal to 10^{-24} cm². The probability of a reaction occurring is therefore expressed as an effective nuclear area through which the incident particle must pass for the reaction to take place. The particle flux (ϕ) is a measure of the number of particles available to take part in the reaction. In the case of neutrons, for example, it is expressed as neutrons/cm²-second, and is the number of neutrons which pass through a plane of area 1 cm² in one second. Since the number of target atoms changed in a

Radioisotopes and Labelled Compounds

nuclear reaction is usually extremely small (less than 1 in 10^6) the number of target atoms, N_{tg}, remains essentially constant. The cross-section, σ, is also constant for any particular reaction, so that for any reaction in which the particle flux, ϕ, is constant the rate at which radioactive atoms are produced will be constant.

The activity of radioisotopes produced by nuclear reactions

The quantity of a radioisotope is usually specified in terms of its *activity*, or disintegration rate. The units of activity most frequently used are the *curie* (Ci) and its sub-multiples the millicurie (mCi), 10^{-3} Ci; the microcurie (μCi), 10^{-6} Ci; and the picocurie ($\mu\mu$Ci or pCi), 10^{-12} Ci. The curie is defined as that quantity of a radioisotope which has a disintegration rate of $3\cdot7 \times 10^{10}$ disintegrations per second, and it is almost equal to the disintegration rate of 1 g of radium-226.

The radioactive decay law (which is discussed more fully in Experiment 1) relates the number of radioactive atoms, N_r, in a radioactive material to their decay rate, dN_r/dt, by the expression

$$\frac{dN_r}{dt} = -\lambda N_r \tag{2.7}$$

where λ is the radioactive decay constant for that radioisotope. We have seen above that the rate of production of radioactive atoms by a nuclear reaction is constant. Initially the decay rate of an isotope being formed will be zero, but this will gradually increase as the number of atoms formed by the reaction increases. The change occurring in the number of atoms present in the time interval dt will be given by the difference between the production rate of the isotope and its decay rate, namely

$$dN_r = (N_{tg}\sigma\phi - \lambda N_r)\, dt \tag{2.8}$$

The solution of this equation enables the number of atoms or the activity of the isotope at any time t to be calculated, thus:

$$N_r = \frac{N_{tg}\sigma\phi}{\lambda}(1 - e^{-\lambda t}) \tag{2.9}$$

or

$$\frac{dN_r}{dt} = -\lambda N_r = -N_{tg}\sigma\phi(1 - e^{-\lambda t})$$

where λ is the radioactive decay constant of the isotope produced.

The quantity $(1 - e^{-\lambda t})$, known as the *growth factor*, approaches unity as the time for which the nuclear reaction proceeds, called the *irradiation time*, is increased to several times the half-life of the radioisotope formed. Since $\lambda t = 0.693 t/t_{1/2}$, where $t_{1/2}$ is the half-life of the radioisotope, it is useful to evaluate the growth factor for various multiples of $t/t_{1/2}$ as in Table 1. At saturation, that is when

Table 1 Values of growth factor

$t/t_{1/2}$	$1 - e^{-\lambda t}$
0·1	0·067
1·0	0·500
2·0	0·750
3·0	0·870
5·0	0·969
7·0	0·992
12·0	0·9998

the growth factor is equal to unity, the rate of decay of the radioisotope is just equal to its rate of production, and this will then be the maximum activity which can be produced in the target material. We may note that, after irradiation for only one half-life, the activity produced is already 50% of the saturation activity, and that after irradiating for five half-lives it has reached 97% of saturation.

THE SPECIFIC ACTIVITY OF RADIOISOTOPES

The *specific activity* of a radioactive material is the ratio of its activity to the mass of the element or compound present. This is a most important quantity for the specification of radioactive preparations for practical applications. For example, in following the uptake of sulphate in a plant (see Experiment 29) the use of high-specific-activity labelled sulphate will enable the uptake of very small amounts of sulphate to be followed. The use of a low-specific-activity labelled sulphate, however, would mean that the plants would have to remove comparatively large amounts of inactive sulphate in order to show significant uptake of activity. The sensitivity for following the uptake of sulphate would consequently be greatly reduced. Again, in the determination of the solubilities of slightly soluble salts (see Experiment 17), it is advantageous to use labelled precipitates of high specific activity for the preparation of saturated solutions of the salts. This increases the sensitivity of the method and enables extremely low solubilities to be determined.

A radioisotope preparation which contains only the radioactive isotope itself and no inactive isotopes of the same element is referred to as being *carrier free*. Inactive isotopes of the same element will

Radioisotopes and Labelled Compounds

behave chemically the same as the radioisotope itself and are referred to as *carriers*. The number of atoms present in a given activity of a carrier-free radioisotope may be calculated from the decay law, as in the following example for 1 Ci of phosphorus-32.

^{32}P: half-life = 14·3 days

$$\lambda = 5·61 \times 10^{-7} \text{ s}^{-1}$$

For an activity of 1 Ci, substitution of the decay rate and this value of λ in eqn. (2.7) gives

$$3·7 \times 10^{10} = 5·61 \times 10^{-7} N_r$$

whence

$N_r = 6·59 \times 10^{16}$ atoms of phosphorus per curie

and dividing by the Avogadro number, gives the number of moles of phosphorus per curie as

$$6·59 \times 10^{16} / 6·02 \times 10^{23} = 1·09 \times 10^{-7} \text{ moles of phosphorus per curie}$$
$$= 3·50 \times 10^{-6} \text{ grammes of phosphorus per curie}$$

which is a specific activity of

$2·85 \times 10^5$ Ci per gramme of phosphorus

or

$0·95 \times 10^5$ Ci per gramme of PO_4^{3-}

The results of similar calculations for several radioisotopes are given in Table 2. We may note that the specific activity generally decreases as the half-life of the radioisotope increases. Typically, an

Table 2 Specific activities of some carrier-free radioisotopes

Radioisotope	Half-life	Ci/g of isotope	Grammes of isotope in 1 μCi
Manganese-56	2·6 h	$2·2 \times 10^7$	$4·5 \times 10^{-14}$
Sodium-24	15·0 h	$9·1 \times 10^6$	$1·1 \times 10^{-13}$
Iodine-131	8·0 d	$1·26 \times 10^5$	$7·9 \times 10^{-12}$
Phosphorus-32	14·3 d	$2·8 \times 10^5$	$3·5 \times 10^{-12}$
Calcium-45	160 d	$1·8 \times 10^4$	$5·4 \times 10^{-11}$
Cerium-144	285 d	$3·15 \times 10^3$	$3·17 \times 10^{-10}$
Caesium-137	30 y	83·3	$1·2 \times 10^{-8}$
Carbon-14	5760 y	7·5	$1·34 \times 10^{-7}$

activity of about 1 μCi is used in tracer experiments, and inspection of the table also shows that in all the cases quoted this will correspond to less than 1 μg of material and in some cases to very much less.

SOME EXAMPLES OF RADIOISOTOPE PRODUCTION METHODS

The general aim in radioisotope production is to obtain a product of as high a specific activity as possible. This allows the user the possibility of adjusting the specific activity to suit his own purpose since it is always possible to add inactive carriers to the radioisotope preparation. Carrier-free radioisotopes can generally be obtained only by production methods which yield a product that is different in chemical identity from the target material. The charged-particle reactions, neutron-proton, neutron-alpha and uranium-fission reactions fulfil this criterion. The neutron-capture reaction does not, but in some cases β-decay of the immediate product of neutron capture yields a radioactive product which may then be chemically separated from the original target. The production of iodine-131 by neutron irradiation of tellurium is an example of such a production method. Table 3 summarizes some examples of these radioisotope production reactions. It may be noted that charged-particle bombardment generally leads to electron capture or positron decay and neutron capture to β^--decay of the product isotope.

The preparation of carbon-14-labelled compounds

Carbon-14 has a half-life of 5760 years and it decays by β-particle emission, the maximum energy of the β-particles emitted being 0·167 MeV. The low energy of the β-particles makes it necessary to apply corrections for their absorption within the source when making measurements of carbon-14 activities (see Experiment 16). This low energy of the β-particles, however, is an advantage in the technique of autoradiography (see Experiment 27.2) since the short range of the β-particles in plant tissues, for example, enables a sharp distinction to be made between those tissues which contain carbon-14 and those which do not. The β-particles from labelled areas do not have enough energy to enter adjoining unlabelled tissues. The major advantage of carbon-14 is that it can be used to label a very large number of organic compounds. It is therefore most useful as a tracer for studying reactions in organic chemistry and in biochemical processes.

Carbon-14 is produced by neutron irradiation of nitrogen in the reaction ^{14}N (n, p) ^{14}C. A nitrogen compound, usually aluminium nitride or ammonium nitrate, is subjected to neutron irradiation.

Table 3 Some radioisotope production reactions

Reaction	Decay mode of product
Proton bombardment $^{56}Fe(p, n)^{56}Co$ $^{60}Ni(p, \alpha)^{57}Co$	 E.C. (electron capture) E.C.
Deuteron bombardment $^{52}Cr(d, 2n)^{52}Mn$ $^{10}B(d, n)^{11}C$ $^{24}Mg(d, \alpha)^{22}Na$ $^{56}Fe(d, n)^{57}Co$ $^{56}Fe(d, \alpha)^{54}Mn$	 E.C., β^+ β^+ β^+ E.C. β^+
Fission of uranium-235 $^{235}U(n, f)^{90}Sr(^{90}Y)$ ^{137}Cs ^{131}I $^{144}Ce(^{144}Pr)$	 β^- β^- β^- β^-
Neutron-alpha reactions $^{27}Al(n, \alpha)^{24}Na$ $^{6}Li(n, \alpha)^{3}H$	 β^- β^-
Neutron-proton reactions $^{14}N(n, p)^{14}C$ $^{35}Cl(n, p)^{35}S$ $^{32}S(n, p)^{32}P$	 β^- β^- β^-
Neutron capture reactions $^{238}U(n, \gamma)^{239}U(\beta^-)^{239}Np(\beta^-)^{239}Pu$ $^{130}Te(n, \gamma)^{131}Te(\beta^-)^{131}I$ $^{127}I(n, \gamma)^{128}I$ $^{23}Na(n, \gamma)^{24}Na$ $^{41}K(n, \gamma)^{42}K$	 α β^- β^- β^- β^-

Subsequently, the compound is treated with sulphuric acid and hydrogen peroxide, which releases the carbon-14 in the form of carbon-14-labelled carbon dioxide and methane. The methane is oxidized to carbon dioxide, which is then absorbed in alkali and precipitated as carbon-14-labelled barium carbonate. This labelled carbonate is used for the production of labelled compounds by the following general methods.

Chemical synthesis. It is possible to label only selected atoms in the synthesized molecule by careful control of the synthetic method. Carbon-14-labelled sodium acetate, for example, may be prepared so that it is labelled in either the carboxyl carbon-atom position (carbon-1) or in the terminal carbon-atom position (carbon-2). The preparation of sodium-acetate-1-C14, starting with labelled carbon dioxide, may be carried out by the Grignard reaction

$$CH_3.MgI \xrightarrow{^{14}CO_2} CH_3{}^{14}COOMgI \xrightarrow{H^+} CH_3{}^{14}COOH \xrightarrow{NaOH} CH_3{}^{14}COONa$$

The synthesis of sodium-acetate-2-C14 from labelled carbon dioxide, on the other hand, requires three synthetic steps, as follows:

I $\quad {}^{14}CO_2 \xrightarrow[\text{KCuAl}_2\text{O}_3 \text{ catalyst}]{3H_2 \text{ (250 atm, 285°C for 6 h)}} {}^{14}CH_3OH$

II $\quad {}^{14}CH_3OH \xrightarrow{HI} {}^{14}CH_3I$

III $\quad {}^{14}CH_3I \xrightarrow{KCN} {}^{14}CH_3CN \xrightarrow{NaOH} {}^{14}CH_3COONa$

Biosynthesis. In this method advantage is taken of natural synthetic processes for the synthesis of complex molecules. Specific enzymes may be used to catalyse particular chemical reactions. The process of photosynthesis in plant leaves may be used to synthesize labelled sugars and starch from carbon-14-labelled carbon dioxide (see Experiment 27.3). Micro-organisms such as *Chlorella* may be used to synthesize labelled proteins and amino acids (see Experiment 27.5).

Preparation of tritium-labelled compounds

Tritium, or hydrogen-3, has a half-life of 12·26 years and decays by β-particle emission with a maximum β-particle energy of 0·018 MeV. The low energy of the β-particles makes tritium-labelled compounds impossible to detect with the simple Geiger counters that are used in the experiments described later, although it can be detected when incorporated in tissues by the technique of microscopic autoradiography (see Experiment 30). Since tritium, like carbon-14, may be used to label organic compounds, it is also an important tracer in organic and biochemistry.

Tritium is produced by neutron irradiation of lithium-6, one of the two natural isotopes of lithium, in the reaction ${}^6Li(n, \alpha){}^3H$. Lithium fluoride is generally used for the irradiation, and tritium is released as molecular tritium, ${}^3H_2$, on solution of the irradiated fluoride. Many organic compounds can be readily labelled by the simple technique of exposing them to tritium gas. Hydrogen is a very labile atom, and its exchange for a tritium atom of the tritium gas is induced by the effect of the tritium β-particles on the chemical bonding in the molecule to be labelled. The method is not selective for labelling particular hydrogen atoms, but it avoids complex and often expensive chemical syntheses.

SPECIFICATION AND SPECIFIC ACTIVITY OF LABELLED COMPOUNDS

A labelled compound, say a carbon-14-labelled compound, may be labelled specifically, uniformly or generally. In a specifically labelled compound only one or more specified atoms are labelled. A compound which is uniformly (U) labelled does not have any particular atom labelled, so that each carbon atom of the molecule has the same probability of being a carbon-14 atom. If the compound is generally (G) labelled then each carbon atom position is again labelled but not all the carbon atoms are labelled to the same extent.

The description of a labelled organic compound should include the nature of the labelled atom, the position of the labelled atom and the specific activity of the material. For example, D-glucose is obtainable labelled with carbon-14 in the following ways:

D-glucose-C14(U)	Each carbon atom is labelled, not necessarily in the same molecule
D-glucose-1-C14	Carbon atom-1 is labelled
D-glucose-2-C14	Carbon atom-2 is labelled
D-glucose-6-C14	Carbon atom-6 is labelled

D-glucose-1-C14, D-glucose-2-C14 and D-glucose-6-C14 may be obtained with specific activities of up to 40 mCi/mM and D-glucose-C-14(U) with a specific activity of up to 240 mCi/mM. It may be noted that since there are six carbon atoms in the molecule, the specific activity of 240 mCi/mM for D-glucose-C14(U) is exactly equivalent to a specific activity of 40 mCi/mM for D-glucose-1-C14 so far as labelling of carbon atom 1 is concerned. For experiments in which only carbon atom 1 needs to be labelled the presence of labelling in other positions is of no advantage and may make the interpretation of results more difficult.

The quantity of a labelled compound is usually specified by its activity. Thus, if 0·05 mCi of D-glucose-C14(U) is purchased, of specific activity 150 mCi/mM, then the actual weight of glucose obtained is

$$0·05/150 = 0·00033 \text{ mM}$$
$$= \frac{0·00033 \times 180}{1000} \text{ g of glucose}$$
$$= 6·0 \times 10^{-5} \text{ g of glucose}$$

A similar calculation shows that 0·05 mCi of D-glucose-1-C14 of specific activity 30 mCi/mM would correspond to $3·0 \times 10^{-4}$ g of glucose.

Tritium-labelled compounds are generally available with much higher specific activities than are carbon-14-labelled compounds, typically in the range 100–5000 mCi/mM. For example, D-glucose-6-T is available with a specific activity of 1000 mCi/mM, and 0·05 mCi would correspond to 10^{-5} g of glucose.

PURITY OF LABELLED MATERIALS

There are three purity criteria which may be applied to a labelled substance. *Chemical purity* is the percentage of the material present in the specified chemical form, irrespective of any labelling; *radioisotope purity* is the percentage of the total radioactivity present as the specified radionuclide; and *radiochemical purity* is the percentage of the radionuclide present in the specified chemical form and in the specified labelled position.

BEHAVIOUR OF MATERIALS OF HIGH SPECIFIC ACTIVITY

The activities which are required for tracer experiments, as has been shown above, correspond to extremely small masses of material when high-specific-activity materials are used. The behaviour of such small masses of material often appears different from that of larger amounts of the corresponding inactive chemical. The chemical behaviour of the radioactive material is not of course any different from that of very small masses of inactive material, but the fact that radioactivity allows work to be carried out with such small quantities permits observation of properties that would otherwise go unobserved.

With carrier-free solutions of inorganic ions, it is possible that a significant fraction of the solute may be adsorbed upon the surfaces of the containing vessel, or upon non-isomorphous precipitates formed during chemical processing, or upon silicon or dust particles present in the solution. Such adsorption is important, for example, in the case of phosphorus-32-labelled phosphate in neutral or alkaline solutions where adsorption upon glass surfaces can result in a "loss" of the active material. The addition of a small amount of inactive "hold-back" carrier, or in the case of phosphate the addition of hydrochloric acid to make the pH 2-3, will prevent such behaviour. Carrier-free labelled phosphate and many labelled organic compounds are also liable to be metabolized by micro-organisms. In the examples discussed above, 0·05 mCi of labelled glucose was seen to correspond to only 10^{-4} to 10^{-6} g of glucose, and this is about 100 times the quantity that may be used in some tracer experiments. Metabolism by micro-organisms contaminating the apparatus could obviously cause the degradation of most of the material, so that initial sterilization of the apparatus and the use of aseptic techniques might be necessary.

Another characteristic of very dilute solutions of carrier-free materials is their tendency to form *radiocolloids*. These are not always true colloids but may simulate colloidal behaviour because of adsorption of the active species on colloidal particles of foreign matter in the solvent. In some cases slightly soluble compounds may be colloidally dispersed even though the solubility product for the compound is not exceeded. This occurs, for example, when lead-212 and bismuth-212, both of which form slightly soluble hydroxides, are present in carrier-free form in ammonium hydroxide solution. The colloidal behaviour of lead/bismuth solutions is suppressed by the presence of hydrochloric acid. The addition of reagents which form soluble complexes, for example citrate with the lead ion, also reduces the extent of radiocolloid formation. Sedimentation occurs in such radiocolloid solutions and can lead to vertical concentration gradients in solutions which have been left standing even for only a few days. It has been reported that as much as 98% of bismuth-212 activity may be removed from a carrier-free solution by centrifuging in an ordinary laboratory centrifuge for half an hour. Careful removal of particulate material from distilled water can sometimes reduce the tendency to radiocolloid formation, but whenever possible it is advisable to add sufficient carrier at all stages of an experiment to ensure that the system follows the macro-scale chemical behaviour. The amount of carrier necessary in most cases is not more than 100 μg.

RADIATION INDUCED DECOMPOSITION

Radiation-induced decomposition is another effect which can cause the radiochemical purity of a labelled material to change with time. Energy absorption by the sample from the radiations emitted in radioactive decay can lead to chemical decomposition. If a labelled atom, carbon-14 for example, in a molecule undergoes radioactive decay, it will emit a nuclear β-particle and the recoil nucleus will have an energy which is just about sufficient to disrupt the molecule. This primary event, however, results in only one molecule being destroyed, the process cannot be controlled in any way, and the overall effect on the number of molecules present is negligible. The β-particle released, however, with an energy of 167 keV compared to an energy of only a few electronvolts required to break a chemical bond, is capable of completely disrupting a few thousand molecules. The absorption of radiation by the remaining labelled molecules is the primary radiation absorption event. It may be followed by secondary chemical reactions of the resulting damaged molecules.

The effect of the primary absorption event may be reduced in a number of ways, for example

1. By dilution of the active material with a carrier—this results in a lowering of the specific activity.
2. By dispersion of the active material in a diluent from which it may subsequently be separated.
3. By geometrical distribution of the active material so that the nuclear radiations are not absorbed by the labelled material.

The addition of carrier material is probably the most effective way of reducing the radiation-induced decomposition, but the necessary reduction in the specific activity of the material is not always permissible. One of the most effective ways of adding a diluent is to dissolve the compound in a solvent. The solvent molecules themselves will be affected by the absorption of radiations. Water and hydroxylic solvents, for example, produce reactive hydrogen atoms, hydroxyl radicals and hydrogen peroxide. These reactive species may destroy the solute which is to be protected. Aromatic solvents, such as benzene, are more resistant to radiation decomposition than aliphatic and hydroxylic solvents, forming ultimately a non-reactive polymeric material, and may with advantage be used as diluents. The only solid diluent which has been widely used is filter paper, and its effect may be partly due to the geometrical distribution of the active material. The "geometry" method is most successful for the more energetic radiations; for example, in the storage of vitamin-B_{12} which has been labelled with cobalt radioisotopes, the material may be deposited as a thin layer on the inside of a glass ampoule by freeze drying of a solution. Absorption of the energetic gamma radiation emitted is very unlikely in the thin layer so produced. Whatever method of storage is used, the values quoted for the radiochemical purity of a labelled material cannot be relied upon after a long period of storage.

Chapter 3
The detection of nuclear radiation

The radiations which may be emitted in nuclear decay processes are α-particles, β-particles and electromagnetic radiation. This electromagnetic radiation ranges in energy from the X-rays that are emitted in electron capture to the more energetic γ-rays that result from excited states of nuclei. These radiations have been detected in a variety of ways. The cloud chamber, for example, allows the visualization of α- or β-particle tracks (see Experiment 13 and Plate 10). Alpha particles may also be readily detected with ion chambers and semiconductor detectors. The Geiger counter can be used for the detection of all these kinds of radiation. The operation of any of these detectors depends upon the interaction of the radiations with the material of which the detector is made. This interaction is the same for all the detectors, namely the radiation loses its energy by causing ionization in the detector. To understand the principles on which detectors operate, therefore, we shall consider first the mechanisms by which nuclear radiations interact with matter.

INTERACTION OF NUCLEAR RADIATION WITH MATTER

We shall consider all matter as being composed essentially of atoms. These atoms may be combined to form chemical compounds which may be in the gaseous, liquid or solid state. It is, however, only necessary for us to consider a generalized picture of matter which we will regard as a random collection of atoms as shown in Fig. 3.1. Each of these atoms consists of a relatively very large electron cloud in which the extra-nuclear electrons are to be found, with a very small nucleus at its centre. The overall diameter of the atom is about 100 000 times that of the nucleus. For this reason, it is much more likely that any radiation which passes through an atom will interact with electrons rather than the nucleus. At each interaction or collision with an electron the nuclear radiation will lose energy to the electron. An electron in an atom which gains energy from such an interaction may either be raised to a higher energy level, resulting in an excited atom, or it may be removed completely from the atom,

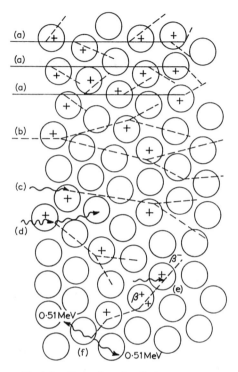

Fig 3.1 Absorption of radiation in matter

(a) Absorption of α-particles
(b) Absorption of β-particles
(c) Photoelectric absorption of γ-rays
(d) Absorption of γ-rays by Compton scattering
(e) Absorption of γ-rays by pair production
(f) Annihilation of a positron with an electron

resulting in ionization with the formation of an *ion pair*. An ion pair consists of a free electron, which may have considerable kinetic energy, and a positively charged atom. The energy required to ionize an argon atom is 15·7 eV, that needed to ionize a nitrogen molecule is 15·6 eV, and that for a bromine molecule, 10·6 eV. The ways in which ion pairs are produced in the absorption of the three kinds of radiation are considered below.

Absorption of α-particles

The α-particle is a helium nucleus with a mass of about 7000 times that of an electron. The laws of classical mechanics apply to collisions between α-particles and electrons, and it may be shown that in these collisions the electrons are scattered through large angles whilst the α-particles continue on a straight-line path. At each collision,

Detection of Nuclear Radiation

kinetic energy is transferred from the α-particle to the electron involved. Since the energy of most α-particles is about 5 MeV, the electron receives sufficient energy to remove it from the atom, and the atom is left positively charged, so that an ion pair is formed at each interaction. This process continues until the α-particles have lost all their energy. Since α-particles are mono-energetic and follow a straight-line path, all α-particles lose their kinetic energy completely after they have covered the same distance, the so-called *range* of the α-particle. The absorption of α-particles is illustrated schematically in Fig. 3.1(*a*).

Absorption of β-particles

The nuclear β-particle is an energetic electron, and when it collides with the electrons in an absorber it is deflected by large angles since this is a collision between particles of equal mass. Its kinetic energy is divided between the particles involved. The energy of β-particles varies from a few hundred electronvolts to several million, so that they always have enough energy to ionize atoms on electron collision. Following collision, the scattered β-particle and electron have sufficient energy to cause further ionization of atoms. Since β-particles have a broad energy spectrum and may be scattered through large angles when they collide, it is not possible to define their range precisely. The overall effect of the absorption is the same as for the α-particle, namely the production of a number of ion pairs. The process is illustrated in Fig. 3.1(*b*).

The positron, or positively charged electron, is absorbed in matter by exactly the same processes as the β-particle. However, whereas a scattered β-particle and an electron are indistinguishable since they have identical charges, the positive charge of the positron allows its path in matter to be distinguished from that of the electrons which it scatters. When the positron has lost all its kinetic energy by collision, it undergoes annihilation with an electron. The positron and an electron disappear and two γ-ray photons, each of 0·51 MeV, are emitted in opposite directions at the point where the annihilation took place. This phenomenon is shown in Fig. 3.1(*f*). Matter is converted to energy in the form of the two γ-ray photons produced, according to the Einstein relationship between mass and energy, $E = mc^2$.

Absorption of γ-rays

The energy of nuclear γ-rays is always much greater than that needed to ionize atoms, so their absorption in matter produces many ion pairs. There are three ionizing processes involved in γ-ray absorption, namely the *photoelectric effect*, *Compton scattering* and *pair production*.

In the photoelectric effect the γ-ray photon transfers all its energy to an inner orbital electron of an atom. This electron is ejected from the atom with an energy equal to that of the incident photon minus the electron's binding energy to the nucleus, and the γ-ray disappears. The electron is then absorbed by the same processes as described for a β-particle (Fig. 3.1(c)).

The Compton scattering process is shown in Fig. 3.1(d). This occurs when a photon interacts with an outer orbital electron in such a way that it transfers only a fraction of its energy to the electron. Both the electron and the γ-ray undergo angular scattering, and the energy of the scattered γ-ray is less than that of the incident γ-ray by an amount equal to the kinetic energy imparted to the electron.

The energies of the scattered and incident γ-rays and the angle of scattering, θ, are related by the equation

$$E_\gamma \text{ (scattered)} = \frac{E_\gamma}{1 + E_\gamma/m_0 c^2 (1 - \cos \theta)}$$

where c is the velocity of light and $m_0 c^2$ is the energy equivalent of an electron mass according to Einstein's mass-energy relationship. It may be noted that E_γ (scattered) is a minimum when $\cos \theta = -1$, i.e. when $\theta = 180°$, or the photon is scattered back along its incident path.

The scattered electron is again absorbed in the same manner as a β-particle

The third γ-ray absorption process is pair production, which takes place in the high electric field close to the nucleus of an atom. A γ-photon of energy greater than 1·02 MeV can interact in this high field to produce an electron-positron pair. This may be regarded as the conversion of energy to matter according to the Einstein equation, and the threshold photon energy necessary for pair production is the energy equivalent of the mass of two electrons. Photon energy in excess of 1·02 MeV appears as kinetic energy of the positron and the electron although it is not always shared equally between them. The positron-electron pair so created is then absorbed in the way described above (Figs. 3.1(e) and (f)).

GAS IONIZATION RADIATION DETECTORS

When a charged particle, an X-ray photon or a γ-ray photon passes through a gas it causes ionization of the gas molecules or atoms in the processes described above. The number of ion pairs produced per unit length of path by the incident radiation is called the *specific ionization*, and approximate values for the radiations we are

Detection of Nuclear Radiation

concerned with are

Radiation	Specific ionization
α-particles	$(5-10) \times 10^4$ ion pairs/cm
β-particles	$(5-10) \times 10^2$ ion pairs/cm
X-ray photons γ-ray photons	Less than 10 ion pairs/cm

If a pair of electrodes are placed in a gas and a potential difference is applied across them then the electrons and positive ions formed in the gas when radiation passes through it will travel in the electric

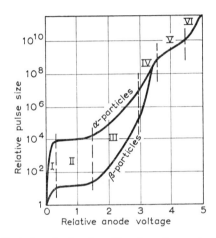

Fig 3.2 Variation of pulse size with voltage for gas ionization detectors

field towards the anode and cathode respectively. If the total number of ion pairs collected at the electrodes for each incident particle or photon is measured as a function of the potential difference between the electrodes, a curve similar to that in Fig. 3.2 is obtained. We shall now discuss the phenomena which correspond to the various regions of this curve.

Region I
In this region the electric field is so small that some of the ion pairs formed recombine due to collisions before they can be collected at the electrodes.

Region II (ionization chamber region)
Here the electric field is just large enough for all the ion pairs formed to be collected before any have time to recombine.

Region III (proportional counter region)

The electric field is here large enough to cause secondary ionization. The electrons of the primary ion pairs are accelerated towards the anode, and when they have gained sufficient kinetic energy they cause ionization of further gas atoms or molecules by collision. In turn the secondary electrons so produced are accelerated and cause further ionizations in the gas and so on, resulting in a cascade, or Townsend *avalanche*, of electrons. The number of ion pairs produced in this way per primary ion pair caused by the radiation is called the *gas amplification factor*, and it increases with increasing voltage. In this region it varies from 1 up to about 10 000. For a given voltage, the number of ion pairs collected at the electrodes, or *pulse size*, is proportional to the number of primary ion pairs caused by the radiation, and for particles which lose all their energy within the gas (particles whose range is less than the counter dimensions) the pulse size will be proportional to the energy of the particle which produced it. It is thus possible, for example, to distinguish between α-particles of different energies. For radiations whose range is greater than the counter dimensions, the pulse size will be proportional to the specific ionization of the radiation which produced it, and it will be possible to distinguish between radiations which have different specific ionizations, for example β-particles and γ-rays.

Region IV (region of limited proportionality)

In this region there is a gradual transition between the characteristics of region III and those of region IV. The pulse size is no longer directly proportional to the number of primary ion pairs produced.

Region V (Geiger–Müller counter region)

At the voltages applied in this region the gas amplification factor is very great, probably between 100 million and 1000 million, and the pulse size is dependent on counter dimensions instead of the primary ionization produced. Thus it is not possible to distinguish between radiations of different energies or specific ionizations.

Region VI (breakdown region)

Here the electric field is so great that many spurious pulses occur and eventually the counter breaks into a continuous discharge which can only be stopped by reducing the applied voltage.

There are three gas ionization detectors which are widely used in elementary work, namely the *ionization chamber*, the *Geiger–Müller counter* and the *spark counter*. We shall discuss the principles and operation of these detectors in this order.

Detection of Nuclear Radiation

The ionization chamber (Region II)

Ionization chambers operate by collecting the primary ions formed when ionizing radiation passes through a gas. The positive and negative ions formed drift towards the electrodes of opposite sign and cause a total charge transference between the electrodes equal to Ne, where e is the charge of the electron and N is the total number of ion pairs formed. If V is the voltage applied between the electrodes and C is the capacitance of the ion chamber, this charge transference will cause a reduction of the voltage between the electrodes of

$$dV = Ne/C$$

If the number of ion pairs, N, is large, as it is with radiations of high specific ionization such as α-particles, then it is possible to detect this drop in the voltage between the electrodes, after amplification, as a pulse. For radiations of lower specific ionization, however, this is not possible and ion chambers are used as integrating instruments which measure the current resulting from the passage of a large number of ionizing particles or photons through the chamber. If the rate at which particles or photons enter the chamber is r per second then the current in the ion chamber, if all the ion pairs formed are collected, is given by rNe. An example of the calculation of the current in an ionization chamber is given in the introduction to Experiment 3. Even with α-particles, the pulse size in an ion chamber due to a single particle is very small and it is necessary to use a high-gain amplifier for pulse counting. For teaching purposes, therefore, ion chambers are generally used in the integrating mode, the ion current being measured with either a pulse electroscope or a high-gain d.c. amplifier.

The Geiger–Müller counter (Region V)

In an electron avalanche such as occurs in the Geiger-Müller region, electrons which are travelling towards the anode can cause excitation of the atoms or molecules of the filling gas of the counter by collision processes. Such excitation occurs when an electron in an atom or molecule is raised to an energy level higher than the ground state. When the excited electron returns to the ground state an ultraviolet photon is emitted, and such photons can subsequently cause release of photoelectrons from the cathode metal or the filling gas. The release of photoelectrons is energetically possible if the photon energy is greater than the work function of the metal (about 4 eV), and in fact this is always so. Photo-ionization of the filling gas requires rather more energy since the ionization potentials of filling gases are generally greater than 11 eV. Electrons will be released by either of these processes in regions of the counter which are remote from the primary avalanche and will be accelerated towards the anode, resulting in secondary avalanches.

A secondary avalanche will be initiated in such a manner, if

$$N_n \varepsilon > 1$$

where N_n is the number of electrons in the nth avalanche, and ε is the probability that an electron in an avalanche will initiate a secondary avalanche in the manner described.

This process of successive avalanches will be terminated when, after the nth avalanche, the increased number of slowly moving positive ions around the anode has reduced the electric field, and hence the gas amplification factor, to the point where $N_n \varepsilon$ is less than unity. The value of ε is around 10^{-5}, so that for the propagation of the avalanches N_n must be greater than 10^5. Electrons accelerated in a given electric field, because of their relatively very small mass, will travel very many times faster than positive ions. All the electrons produced in a series of avalanches are, in fact, collected at the anode in a fraction of a microsecond, but the slow-moving positive ions require some hundreds of microseconds for complete collection at the cathode. The total number of electrons collected in a series of avalanches determines the pulse size from the counter. If the anode voltage is increased, a greater number of avalanches is required to build up a positive-ion sheath around the anode large enough to terminate the series so that the total number of electrons collected becomes greater and the pulse size is increased, as indicated by the slope of the Geiger-Müller region in Fig. 3.2. The pulse size will thus be a function of both counter dimensions and anode voltage, but it is independent of the primary ionization caused by the radiation, so that it will not be possible to distinguish between radiations of different specific ionizations or energies.

After all the electrons produced in a series of avalanches have been collected as described above, the relatively slow-moving positive-ion sheath remains around the anode and is accelerated towards the cathode, where the following processes take place on arrival. The positive ion, which has a potential energy equal to its ionization potential (15·7 eV for argon), on arrival at the cathode is neutralized by withdrawing an electron from the metal. In order to do this the ion must expend energy equal to the work function of the metal. The neutralized atom so formed is generally in an excited state, having an excitation energy equal to the difference between the work function of the cathode metal and the ionization energy of the atom. When the atom returns to the ground state its excitation energy is radiated as an ultraviolet photon which may subsequently release a photoelectron from the cathode. Especially in very high fields, the neutralized atom may collide with the cathode before it has had time

Detection of Nuclear Radiation

to radiate its excitation energy and electrons may then be liberated directly from the cathode. Electrons originating by these processes can initiate a further series of avalanches, causing a secondary pulse. The methods by which these secondary pulses can be prevented, or quenched, are described below.

External quenching. The very rapid fall of potential on the counter anode, which follows the series of electron avalanches, is detected by a sensitive trigger circuit which causes the voltage on the anode to drop immediately to well below the Geiger–Müller region, usually by about 200 V. The anode is held at this reduced voltage for a few hundred microseconds before being restored to the normal operating potential, and during this time the positive ions are collected at the cathode. Any electrons released at the cathode cannot now produce a secondary pulse as the gas amplification factor at the reduced field is too low for avalanche propagation.

Internal or self-quenching. Secondary pulses may also be prevented by the addition of polyatomic molecules, such as ethyl alcohol or the halogens, to the filling gas of the counter. These additives have a lower ionization potential than the filling gas, so that on collision between their molecules and the ionized atoms of the filling gas, electron transfer readily takes place. Each ion of the filling gas makes a large number of collisions on its way to the cathode, many of them with quenching gas molecules, so that electron transfer takes place before such ions can reach the cathode and only ionized molecules of the quenching agent arrive there. On neutralization the quenching gas molecules are left in an excited state, their excitation energy being well above their dissociation energy. The time required for such molecules to dissociate is about 10^{-13} s, whereas the lifetime of the excited state is about 10^{-8} s, so that the excited molecules will dissociate before they can radiate their excitation energy as photons which would, of course, release electrons from the cathode and initiate secondary pulses. In the case of polyatomic organic molecules recombination of the dissociation fragments cannot occur, and the lifetime of the counter is limited by the amount of quenching gas present. Halogen molecules, however, dissociate into halogen atoms which subsequently recombine, so that the life of a halogen-quenched counter is not limited by the amount of halogen present and only very small amounts of the quenching gas need be added.

The quenching mechanism for a halogen-quenched counter is:
1. Radiation interaction by ionization processes: $\rightarrow Ar^+$
2. Electron transfer $\quad Ar^+ + Br_2 \rightarrow Ar + Br_2^+$
3. Discharge at cathode $\quad Br_2^+ + e^- \rightarrow Br_2^*$ (excited molecule)
4. Dissociation $\quad\quad\quad\quad\quad Br_2^* \rightarrow Br + Br$
5. Recombination $\quad\quad\quad\; Br + Br \rightarrow Br_2$

The Geiger–Müller discharge in low-voltage halogen-quenched counters. In pure argon, electrons which, after several collisions, have gained a kinetic energy of about 12 eV will be able to excite argon atoms to the metastable or first excitation level of argon. To cause ionization of an argon atom an electron must acquire a kinetic energy which is at least equal to the ionization potential of argon (15·7 eV). Since electrons can lose their energy to the metastable level, it will only be possible to ionize argon atoms if the field is high enough to ensure that the electrons gain kinetic energies greater than 15·7 eV between two consecutive collisions. If, however, a counter is filled with neon containing only a small amount, about 0·1%, of argon it becomes possible to ionize argon atoms at much lower electric fields by the process described below.

The neon atom has a metastable state and a first excitation level at 16·6 eV and 16·8 eV respectively. At this concentration of neon an electron accelerated by an electric field will make collisions predominantly with neon atoms and it will undergo elastic collisions until it has acquired a kinetic energy greater than 16·6 eV, when it can take part in an inelastic collision, raising a neon atom to the metastable state. An atom in a metastable state has a long lifetime, often of the order of 0·1 s, during which it will make many collisions in which it may transfer its excitation energy to another atom. Since the energy of the metastable level for neon is greater than the ionization potential of an argon atom, a collision between an excited neon atom and an argon atom can result in ionization of the argon atom. These processes are summarized below.

1. Ionization of neon by radiation $\rightarrow Ne^+ + e^-$
2. Electron acceleration until $e^- + Ne \rightarrow Ne^* + e^-$
3. Excited neon/argon collision $Ne^* + Ar \rightarrow Ar^+ + Ne + e^-$

Counters which operate by the above mechanism require electric fields much smaller than those which cause avalanches in pure argon. These counters are halogen quenched, and a typical filling gas consists of 99·8% neon, 0·1% argon and 0·1% bromine. The small number of halogen molecules present do not appreciably absorb the ultra-violet photons which are emitted by those neon atoms which are excited to the first excitation level, and the avalanches are propagated by these photons causing photoelectric emission at the cathode. The Mullard MX168 counter, for example, has this gas filling and operates at low anode voltages because the avalanches are propagated in the manner described above.

Pulse shapes from Geiger–Müller counters. Fig. 3.3 shows the basic input circuit used with Geiger–Müller counters. The operating potential is applied to the anode through the resistor R_1 and the output pulse is taken from the lower end of this resistor. When the entry of ionizing radiation initiates a pulse, the electrons are collected at the anode and the voltage at the lower end of R_1 decreases. This negative-going voltage pulse is fed to the quenching circuit or scaler via the capacitor, C_1, which serves to isolate the anode voltage supply from the succeeding parts of the circuit. The rate at which the voltage at the lower end of R_1 is restored determines

Detection of Nuclear Radiation

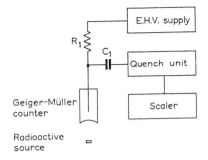

Fig 3.3 Block diagram of the instrumentation used with a Geiger–Müller counter

the pulse shape and is a function of both the mobility of the positive ions and the time constant, RC, of the input circuit. The variation of the pulse shape with this time constant is shown for a typical Geiger–Müller counter in Fig. 3.4. If the time constant is very long compared with the collection time for the positive ions, the rapid fall of the anode potential can be observed together with the further fall which occurs as the positive ions are collected. If the time constant is short compared to the collection time for the positive ions, then the anode potential is restored much more rapidly than the rate at which the positive ions are collected so that only the voltage drop due to the collection of the electrons is observed. The input time constant is therefore always made much less than the collection time for the positive ions, i.e. less than a few hundred microseconds, so that a pulse with a very sharp fall and rapid rise of potential is obtained. The capacitance, C_g, is the sum of the capacitance of the counter and the stray capacitance of the input circuit and is generally about 30 pF, so that for an input time constant of 30 μs the anode load resistance, R_1, should be 1 MΩ.

Dead time, recovery time and paralysis time. The behaviour of the anode voltage on a Geiger–Müller counter, with a fixed input-circuit time constant, is shown in Fig. 3.5. The voltage drops very rapidly as the electrons are collected at the anode, and the electric

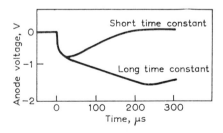

Fig 3.4 Variation of the output pulse-shape of a Geiger–Müller counter with input-circuit time constant

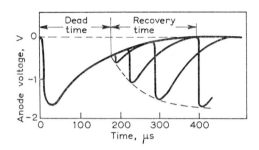

Fig 3.5 Variation of the anode voltage with time for an internally quenched Geiger–Müller counter

field is immediately much reduced by the presence of the positive-ion sheath around the anode, after the entry of ionizing radiation. A second ionizing particle or photon entering the counter cannot initiate a series of avalanches until the field has recovered somewhat, i.e. until some of the positive ions have been collected. The minimum time interval which must elapse before a second ionizing event in the counter can produce a detectable pulse is called the *counter dead time*. The size of the pulse produced by this second ionizing event is dependent upon the time which has elapsed since the first ionizing event, and the counter *recovery time* is defined as the minimum time interval which must elapse between the end of the counter dead time following an ionizing event and the time at which a second ionizing event will cause a pulse of the same size as the first. The time dependence of this pulse size is indicated by the series of pulses shown in Fig. 3.5. Both the dead time and the recovery time vary from one pulse to another, even for the same counter. Typical values are 60 to 150 μs for the dead time and 200 to 300 μs for the recovery time.

An externally quenched Geiger–Müller counter has a square wave of fixed duration applied to the counter anode by the quenching circuit, as shown in Fig. 3.6. The counter is then inoperative for the

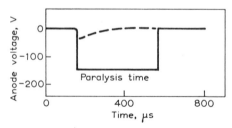

Fig 3.6 Variation of the anode voltage with time for an externally quenched Geiger–Müller counter

Detection of Nuclear Radiation

duration of this square wave, which is called the *paralysis time*, instead of just for the dead time. The paralysis time must be made greater than the combined dead time and recovery time.

Ionizing radiations entering a Geiger–Müller counter during inoperative periods such as the dead time or paralysis time will not contribute to the recorded counting rate. A correction for the counts lost during inoperative periods may be calculated as follows:

Let τ seconds = Inoperative period of the counter
n counts/second = True counting rate
m counts/second = Observed counting rate

Then

True counting rate = Observed counting rate
+ (True counting rate × Total inoperative time)

i.e.

$$n = m + nm\tau$$

from which

$$n = m/(1 - m\tau)$$

For an externally quenched counter τ is the paralysis time and is pre-set at some fixed value, 400 μs being commonly used, and corrections for lost counts can therefore be made accurately. For an internally quenched counter τ is the counter dead time, and an average value must be obtained for it experimentally. The dead time will vary for different counters of the same type and may depend upon the age of the counter and upon the counting rate. Even with internally quenched counters it is usual to apply external quenching as well so that the counter is inoperative for a fixed and known time after each count.

Types of Geiger–Müller counter. Several types of counter to be used in the experiments which follow are described below. Fig. 3.7(*a*) is a line drawing of an end-window counter, which consists of a cylindrical metal cathode with a coaxial anode wire, insulated from the cathode. The end of the cylinder is closed by a thin window of either aluminium or mica. End-window counters are used mainly for β-particle counting, but with very thin mica windows they can also be used for α-particle counting. The filling gas must be nearly at atmospheric pressure to avoid rupturing the thin windows used. The counters are also sensitive to γ-rays, but the efficiency for γ-ray detection is only about 1 % of that for β-particle detection. Because of absorption by the counter window the efficiency is also low for detection of β-particles of low energy such as those from carbon-14 or sulphur-35.

The counter is usually placed in a thick-walled lead *castle* to reduce its response to the natural background, but this is not necessary for the experiments described later. A Perspex castle incorporating a system of shelves to support the sample under the counter window is much cheaper and is quite adequate since it is

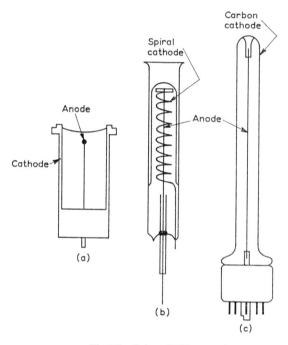

Fig 3.7 Geiger–Müller counters

(a) Mica end-window
(b) Liquid sample counter
(c) Thin glass walled β-counter

not necessary to work at low background levels (see Plates 1 and 2). Such a counter arrangement is used for counting samples in solid form, these being prepared on small metal trays or *planchettes*.

Since the sample emits particles in all directions the counting rate will depend upon the solid angle subtended by the counter window at the sample. It is therefore important, if several samples are to be counted and their activities compared, that they should all be counted in exactly the same position in the castle, i.e. under the same conditions of counter geometry.

Detection of Nuclear Radiation

Radioisotopes in solution can be conveniently counted in counters such as that shown in Fig. 3.7(*b*). The anode wire and a spiral-wire cathode are placed inside a glass tube with sufficiently thin walls to allow energetic β-particles to enter the counter. The liquid sample is contained in an annulus between the counter itself and an outer thick-walled glass tube which is sealed to the counter. The counter is placed in a lead or Perspex castle in which contacts are made between a metal plate and the cathode and between a mercury pool or spring contact and the anode (see Plate 3). This arrangement facilitates the removal of the counter for emptying, cleaning and filling.

Thin glass-walled counters of the type shown in Fig. 3.7(*c*) may be partially immersed in liquids to assay their activity. In these counters the cathode is either a spiral wire or a graphite coating on the inside of the glass wall. Counters of this type are often used in contamination monitors.

The response of Geiger–Müller counters to γ-rays can be enhanced by making the walls of a material of high atomic number, such as lead or tungsten. The resulting absorption of γ-rays in these materials scatters electrons into the counter, and the efficiency for electron detection being high, the overall efficiency for γ-ray detection is improved.

Instrumentation for Geiger–Müller counters. A block diagram of the instrumentation normally used with a Geiger–Müller counter is shown in Fig. 3.3. The power unit provides the counter anode voltage. The stability of this supply need not be very high as the counting rate for a Geiger–Müller counter does not vary greatly with anode voltage at its operating potential. A stability of $\pm 0.1\%$ change in anode voltage for a $\pm 10\%$ change in mains supply voltage is more than adequate. The quench unit, if one is used, imposes a fixed paralysis time on the counter immediately after each particle or photon interacts within it. This enables accurate corrections for lost counts to be made as described earlier. The scaler totals the number of pulses received in a measured time interval. Most scalers can accept positive pulses of amplitude greater than about 5 V or negative pulses of amplitude greater than about 0.2 V.

The e.h.v. supply and the scaler are often incorporated in a single instrument (Plate 1). In order to minimize cost, instruments designed specifically for educational use do not have facilities for the use of quench units (Plate 2). The counting rates to be determined with such equipment should be limited to not greater than 10 000 counts per minute. If a quench unit is not used the inoperative period of the counter will be equal to its natural dead time, which must be determined experimentally (see Experiment 4.2). Corrections for lost counts can be made using the graphs of Appendix 6 if the dead time of the counter is known.

The spark counter (Region VI)

The spark counter operates at the start of the breakdown region of the voltage/pulse-size characteristic. Its use for the detection of α-particles is described in Experiment 6.4 and Fig. E.6.5 is a diagrammatic representation of a spark counter. It consists essentially of a grid of very fine wires separated from an earthed metal plate by a small air gap. The wires are held at a potential with respect to the plate which corresponds to the start of the breakdown region. Under these conditions, if a particle of high specific ionization passes close to a wire, an electron avalanche can be initiated in which the number of electrons multiplies extremely rapidly. The electron energies in the avalanche are high, resulting in atomic and molecular excitations which subsequently lead to optical emission. The result is a visible spark discharge which is maintained by the discharge of the e.h.v. capacitor, C, in Fig. E.6.5. This capacitor is charged through the limiting resistor, R, by an e.h.v. supply. The maximum counting rate of the detector is determined by the value of the time constant, RC, and will generally not be greater than a few thousand counts per minute.

SEMICONDUCTOR DETECTORS

The semiconductor detector may be regarded as a solid-state ionization chamber in which the nuclear radiation loses its energy by absorption in a solid rather than in a gas. As in a gas, energy absorption occurs by the interaction of the radiation with the electrons. In a gas, distinct positive and negative ionic species are formed, but in a solid, the electrons are excited to higher energy states by the radiation and leave vacancies or *holes* in the normal lattice structure of the solid. Both the excited electrons and the holes, which are the equivalent of the positive ions in a gas, are free to wander throughout the solid and act as charge carriers. So that we can discuss the principles of semiconductor detectors more fully, it is necessary to consider first the main features of the band theory of solids.

The band theory of solids

A solid may be regarded as an assembly of nuclei surrounded by electrons, the nuclei being arranged in a regular three-dimensional pattern. Most of the electrons will move in regions influenced by the electric fields due to particular nuclei and so may be regarded as

Detection of Nuclear Radiation

belonging to the particular atoms concerned. Many electrons from the outer shells of the atoms, however, will be influenced by the electric fields which are due to more than one nucleus. These electrons move in the periodic field which arises due to the spatial distribution of the nuclei. The band theory of solids is a detailed study of the behaviour of electrons in such a periodic field, and it reveals that there are certain forbidden energy states which electrons can never occupy (see Fig. 3.8). The behaviour of the electrons in a solid is similar to their behaviour in single atoms in that in both

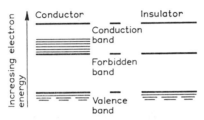

Fig 3.8 Electron distribution in a solid on the basis of the band theory

cases only certain energy states are permissible. In an isolated atom the possible electron energy levels are discrete, whereas for a solid the allowed energies for the electrons fall into rather broad bands. The highest energy states permitted form the *conduction band*, in which the electrons are free to wander throughout the crystal structure. Permitted energy states of lower energy than those in the conduction band form the *valence band* in which the electrons are bound to particular atoms. The conduction band and the valence band are separated by a region of forbidden energy states.

It is now possible, on the basis of this concept, to classify solids as conductors, insulators and semiconductors. A conductor has a filled valence band and a partially populated conduction band; for example, a metal has a half-filled conduction band. An insulator has a filled valence band and an empty conduction band. Since the electrons in the valence band are not mobile in the lattice, there are no charge carriers present and the material is an insulator. The forbidden band gap is about 5–10 eV in the case of an insulator. A semiconductor also has a filled valence band and an empty conduction band, but its *energy gap* is about 1 eV or less, which is within the range of energies that valence electrons may gain thermally, and it is this fact which gives the conductivity of semiconductors a remarkable temperature dependence. The thermal excitation of electrons to the conduction band leaves holes in the valence band, and both these holes and the excited electrons act as charge carriers.

Semiconductor types

A semiconductor, such as pure silicon or pure germanium, which relies solely on thermally excited *electron-hole pairs* for the provision of charge carriers is termed an *intrinsic semiconductor*. It is possible by the controlled introduction of impurity atoms to provide charge carriers which are not dependent upon thermal excitation of electrons. Such semiconductors are termed *extrinsic* or *impurity semiconductors*. Two kinds of impurity semiconductors may be prepared, namely *p-type*, in which there is an excess of positive (hole) charge carriers,

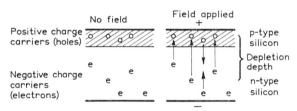

Fig. 3.9 Behaviour of charge carriers in a *p–n* junction

and *n-type*, in which there is an excess of negative (electron) charge carriers. Materials such as silicon and germanium can be used to make *n*-type semiconductors by the addition of phosphorus or arsenic which serve as electron *donors*. The addition of traces of boron or gallium which act as electron *traps*, will cause the formation of holes in the valence band and result in a *p*-type semiconductor.

Semiconductor detectors

In the absence of ionization, the gas in a gas ionization chamber behaves as an insulator. After the entry of nuclear radiation, the resulting ionization makes the gas conducting, and the entry of the radiation is inferred from the observation of the charge transference between the electrodes. Since charge carriers are always present in a semiconductor (even in the highest purity silicon available) the detection of extra charge carriers formed on the absorption of nuclear radiation is not directly possible. It is first necessary to construct a device in which the number of charge carriers can be reduced or depleted to form the counterpart of the nonconducting gas filling of the ionization chamber. Two kinds of semiconductor detectors which have different methods of obtaining charge-carrier-free regions, are described below.

The p–n junction detector. A thin region of *n*-type silicon may be formed on one side of *p*-type silicon, for example by thermal diffusion of phosphorus into the surface. On application of a potential difference to this device, as shown in Fig. 3.9, the holes and electrons drift towards the junction between the two types of semiconductor. If the device is operated with reversed bias (the field polarity shown being reversed) the electrons and holes move away from the junction and a depletion layer in which there are no charge carriers is formed. The thickness of this depletion layer depends upon the applied voltage. The surface of the detector has a layer of evaporated gold to facilitate electrical connections. The nuclear radiation passes through this layer into the depleted region where its absorption produces charge carriers in the form of electron-hole pairs. The collection of these charge carriers

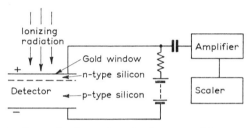

Fig 3.10 Schematic of a *p–n* junction detector and instrumentation

produces a detectable voltage pulse when the counter is used as shown in Fig. 3.10.

Depletion depths from 0·1 to 0·5 mm may be obtained, and since detectors with such small dimensions will have a very low stopping power for γ-ray photons and energetic β-particles, their application is mainly to the detection of α-particles and to a lesser extent, β-particles.

The lithium-drifted detector. This detector consists of *p*-type silicon in which the positive charge carriers (holes) have been compensated by donation of electrons from impurity lithium atoms. The method of preparation involves the thermal diffusion of lithium into the *p*-type silicon under the influence of an electric field until the conductivity of the material falls to nearly zero. Depletion layers as much as a few millimetres thick may be prepared in this way. The presence of the depletion layer is not primarily dependent upon the presence of the electric field. Because of the great thickness of the depletion layer, these detectors are particularly useful for the detection of X-ray and γ-ray photons, since a detector of large volume is necessary for the efficient absorption of photons.

Use of semiconductor detectors

The only semiconductor detectors so far produced sufficiently cheaply for educational use are *p-n* junction devices such as the 20th Century Electronics type SSN 06/E (See Plate 4). This has a sensitive area of 7 mm² and a depletion depth of 0·1 mm as generally supplied for α-particle detection. The evaporated gold window has a negligible stopping power for α-particles, and the range in air may be determined without making any detector window correction.

It is necessary to use an amplifier between the detector and the scaler, so shown in Fig. 3.10, to obtain pulses large enough to operate the scaler. The detector is sensitive to strong light sources and must be used in subdued light. Compared with the thin-end-window Geiger–Müller counter, the semiconductor detector is of little advantage in most of the α-particle experiments described later. However, for Experiment 9, a detector of small dimensions which can be operated in a vacuum is required, and the type SSN 06/E is ideal for the purpose.

SCINTILLATION COUNTERS

A *scintillation counter* detects nuclear radiations by the light emission which they generate on being absorbed by certain materials. The complete counter consists of the material in which the radiation absorption and light emission occurs, namely the phosphor or scintillator, and instrumentation to detect and analyse the intensity of the light emission. If the detector is used solely to determine the counting rate, or activity, of a source the technique is generally known as *scintillation counting*. When it is used to measure the energies involved in nuclear decay processes (e.g. the energies of γ-rays emitted in decay) by analysing the intensity of the light emission from the phosphor, the technique is described as *scintillation spectrometry*.

The scintillation counter can be used for the detection of various kinds of nuclear radiations by using an appropriate phosphor in which the radiation to be counted is absorbed and causes luminescence. Some aspects of the luminescent process, however, are of general application to phosphors and are discussed below. *Luminescence* is the emission of light from a material in which electronic excitations have occurred. These excitations may arise from the absorption of nuclear radiations but can also be caused by such processes as light absorption, chemical reactions, thermal heating and electric discharges. There is a somewhat arbitrary division of luminescent processes into fluorescent events and phosphorescent (or afterglow) events. A *fluorescent event* is one in which the emission of light photons occurs within 10^{-8} s of the excitation. On the other hand, for a *phosphorescent event* the light emission occurs 10^{-8} s or longer after the excitation has occurred and the lifetime of the excited state varies widely, from microseconds to hours, for different materials.

Detection of Nuclear Radiation

The number of light photons, n_t, emitted by a luminescent material in the time elapsed up to t seconds after the excitation occurred is given by

$$n_t = n_\infty(1 - e^{-t/\tau})$$

where n is the total number of photons ultimately emitted and τ is the decay time. Thus it can be seen that τ seconds after the excitation has occurred, the number of photons which will have been emitted is

$$n = n_\infty(1 - e^{-1}) = 0\cdot 63 n_\infty$$

The decay time, τ, may therefore be defined as the time which elapses after excitation for 63% of the total light emission to have occurred. It is therefore a measure of the effectiveness of a luminescent material for the resolution of successive nuclear interactions.

Phosphors for scintillation counting

Scintillation counters may be used for α-particle, β-particle, γ-ray and neutron counting by the selection of appropriate phosphors. Counting efficiencies of nearly 100% can often be obtained, as in γ-ray counting using sodium iodide crystals or in the internal sample counting of β-emitting isotopes such as carbon-14. The mechanism by which the phosphor absorbs the nuclear radiation and re-emits the absorption energy as light varies for the different classes phosphor is discussed below. The broad classes into which phosphors may be divided are the inorganic, organic and solution phosphors. The inorganic phosphors are ionic crystals, for example polycrystalline zinc sulphide for α-particle detection and single crystals of sodium iodide for γ-ray detection. The organic phosphors consist of large molecular crystals such as anthracene and trans-stilbene and are used mainly for β-particle detection. The third class of phosphors, the solution phosphors, includes luminescent materials in solid solution in polymerized material and luminescent materials in liquid solution. Details of a selection of phosphors for various purposes are given in Tables 4 and 5 (overleaf).

The mechanism of energy absorption and emission in inorganic crystalline phosphors. The mechanism of scintillation production in this case may be explained on the basis of the band theory of solids (see page 48). A pure single crystal of an alkali halide has its valence band completely filled and an empty conduction band. The energy levels between these bands are forbidden and there are normally no electrons in them. There are two processes by which electrons may enter these forbidden levels. An electron may be

Table 4 Phosphors for scintillation counting

Description of phosphor	Decay time, μs	Maximum of emission band, nm	Notes
Inorganic scintillators			
NaI(Tl), single crystal of sodium iodide with thallium impurity centres	0·3	420	Used for γ-ray detection and spectroscopy
NaI, pure single crystal of sodium iodide	0·06	303	Used for γ-ray detection and spectroscopy at higher counting rates
ZnS(Ag), polycrystalline zinc sulphide with silver impurity centres	0·04–0·1 and 5·0	450	Used in thin layers for α-particle detection. The powder is not transparent to the emitted light
Organic scintillators			
$C_{14}H_{10}$, single crystals of anthracene	0·03	448	Used for electron and β-particle spectroscopy, and for γ-ray detection at high counting rates. Not as suitable as NaI(Tl) for γ-ray spectroscopy
$C_{14}H_{12}$, single crystal of trans-stilbene	0·003–0·008	384	
$C_{18}H_{14}$, single crystal of p-terphenyl	0·03	391	
Scintillators in solid solution or dispersion			
Polystyrene plastic scintillator, 1·6% w/v tetraphenyl butadiene in polystyrene	0·003	450	Sensitive to α, β and γ radiations. Large volume γ-ray detectors are economic. Also used for β-particle detection and spectroscopy
Polyvinyltoluene plastic scintillator, 3·6% w/v p-terphenyl and 0·09% w/v tetraphenylbutadiene in polyvinyltoluene	0·003	445	
Liquid solution scintillators			
4 g PPO, 0·1 g POPOP, 1000 cm³ toluene (see table 5)	0·003	425	Liquid scintillators are used for internal sample counting of β-active materials, especially sulphur-35, carbon-14 and tritium
50 g naphthalene, 7 g PPO, 0·05 g POPOP, 1000 cm³ p-dioxane	0·003	425	Naphthalene is present to reduce the quenching effect of the dissolved sample. The p-dioxane is for water-soluble samples

Table 5 Some chemicals used in scintillation counting

Name	Abbreviation	Structure
Naphthalene	$C_{10}H_8$	
Anthracene	$C_{14}H_{10}$	
p-terphenyl	TP	
Toluene	C_7H_8	
Xylene	C_8H_{10}	
p-dioxane	$C_4H_8O_2$	
2,5-diphenyloxazole	PPO	
1,4-bis-(2-(5-phenyloxazolyl)-benzene	POPOP	
Vinyl toluene		
Polyvinyl toluene	PVT	

excited to just below the empty conduction band and the configuration is stable provided that the electron remains associated with the hole which it leaves in the valence band. The resulting electron-hole pair, called an *exciton*, is free to move throughout the crystal. Also, if impurity atoms are introduced in the crystal such that the electron levels associated with them fall between the conduction and valence bands of the pure crystal, then at discrete points in the crystal there will be impurity levels, or traps, which can capture electrons that have been excited to the conduction band.

These phenomena are illustrated in Fig. 3.11. In the case of sodium iodide crystals, the impurity added is thallium.

The passage of a γ-ray photon through the crystal may result in the production of a photo-electron or a Compton-scattered electron. In either case the electron energy is likely to be very high and it will be dissipated by electron-electron collisions in the crystal, resulting

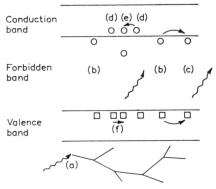

Fig 3.11 Mechanism of excitation and light emission in ionic crystals

A γ-photon is absorbed in the crystal (a) and the scattered electron causes the following phenomena: exciton formation (b), exciton migration and recombination at impurity centres (c); excitation to the conduction band (d), electron mobility in this band and trapping at impurity centres (e); hole migration and de-excitation of electrons in impurity traps (f)

in the excitation of a number of electrons to the conduction band, the creation of a corresponding number of holes in the valence band, and the formation of a number of excitons. The process of de-excitation takes place largely at the impurity centres. Exciton pairs may recombine at an impurity centre, filling the hole in the valence band; or an impurity centre may trap an electron from the conduction band and a hole from the valence band, a photon of energy equal to the excitation energy of the impurity level being emitted. Such photons do not have sufficient energy to excite electrons from the valence band to the conduction band, and the crystal is therefore transparent to this fluorescent radiation. The total number of such light photons emitted is proportional to the energy of the photo-electron or Compton-scattered electron.

The mechanism of energy absorption and emission in solid organic phosphors. In phosphors such as anthracene or trans-stilbene the energy absorption and emission process is essentially one of molecular excitation and de-excitation. The absorption of a γ-ray photon (or a nuclear β-particle) will again cause the production of a high-energy electron whose energy will gradually be degraded until the equivalent number of molecular excitations have been produced. The general features of the energy level diagram for molecular excitation are shown in Fig. 3.12. Electronic excitation requires a few electronvolts of energy and is most likely to take place at the extremities of molecular vibrations. In the excited state

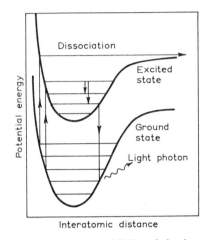

Fig 3.12 **Mechanism of excitation and light emission in molecular crystals**

of the molecule the vibrational energy is generally greater than it was in the ground state. During the lifetime of the excited state this vibrational energy will generally become somewhat degraded due to vibrational transitions and emission in the infra-red, so that on de-excitation the fluorescent photon is of lower energy than that for the initial excitation. Some quenching of the fluorescent radiation may occur due to molecular dissociation on excitation to high vibrational energy levels.

The mechanism of energy absorption and emission in phosphor solutions. These scintillators consist of an organic compound dissolved in a liquid organic solvent or in a polymeric material in the case of plastic phosphors. They are generally used for β-particle counting and spectroscopy. The absorption of a nuclear β-particle leads initially to the excitation of many solvent molecules, since statistically interactions with solvent molecules are most likely because of their great preponderance over the number of solute molecules. The excitation energy of the solvent is then transferred to the solute molecules, probably by processes such as exciton migration in the solution and photon emission and re-absorption. The wavelength of the photons subsequently emitted by excited solute molecules is often not well matched to the sensitive spectral region of photomultipliers, about 400 nm, and so a further solute, called a *wavelength shifter*, is usually added. This solute, POPOP is a commonly used one, absorbs the photons emitted by the excited solute molecules and subsequently emits longer-wavelength fluorescent radiation. The compound whose radioactivity is to be measured is also dissolved in the scintillator solution, and this can result in a serious loss of efficiency of the scintillator, due to absorption of emitted photons or to interference with the mechanism of energy transfer from solvent to solute. It has been found that the addition of relatively large amounts of naphthalene or diphenyl to the scintillator solution can largely inhibit quenching due to the latter, but compounds which have strong absorption bands near 400 nm may be quite unsuitable for scintillation counting in this way. For samples which are soluble in organic solvents, xylene and toluene are generally used as the scintillator solvent. For samples which are soluble in water, the solvent used is p-dioxane, in which about 10% v/v of an aqueous solution may be dispersed.

Detection of the light emitted by the phosphor

The interaction of nuclear radiation in the phosphor causes the emission of light by the process of fluorescence in the phosphor. The light from such an event can be detected with high sensitivity and its intensity may be measured using a photomultiplier. A sectional drawing of a photomultiplier is shown in Fig. 3.13, and its functional parts are discussed below. The light-sensitive material forms the photocathode and is generally deposited on the inside of the glass envelope. This material absorbs the light photons emitted by the phosphor and emits electrons. The number of electrons emitted is proportional to the intensity of the light from the phosphor, which is in turn proportional to the amount of energy deposited in the phosphor by the nuclear event. Caesium antimonide, Cs_3Sb, has been one of the most widely used cathode materials, but the more efficient tri-alkali cathode, KNa_2CsSb, has in recent years tended to displace it. The electrons emitted by the photocathode are focused onto the first-stage electrode, or *dynode*, of the electron multiplier. The electron multiplier consists of a series, usually 10 or 11, of dynodes each of which is held at a potential about $+100$ V above the previous one, the first dynode being held at a positive potential above the photocathode. An electron leaving a dynode is accelerated by the potential difference between that and the next dynode, and on collision with the latter will have sufficient energy to cause the emission of a number of secondary electrons. These electrons are accelerated to the next dynode, where they cause electron multiplication in the same way. Electrons leaving the final dynode of the multiplier are collected at the collector anode. This causes a momentary voltage drop, or pulse, at the anode end of the load resistor R (Fig. 3.13), and this pulse is fed via the blocking capacitor C for amplification and counting.

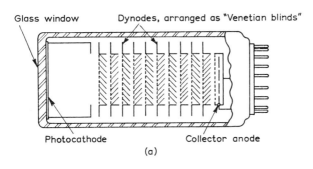

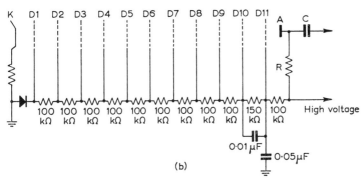

Fig 3.13 Sectional drawing of a photomultiplier and a typical circuit diagram

(a) Photomultiplier
(b) Circuit diagram of dynode resistor chain
A. Collector anode
C. Blocking capacitor
K. Photocathode
R. Load resistor

The electron gain, G, of a photomultiplier with n dynodes is given by the equation

$$G = \delta^n$$

where δ is the average number of secondary electrons released by an electron incident upon a dynode. Typical photomultiplier electron gains are of the order of 10^6, so that the pulse amplitude from the device is 10^6 times that of a simple photocell.

Instrumentation for scintillation counting and spectrometry

A scintillation counter or spectrometer requires instrumentation to supply the various voltages to the dynodes of the photomultiplier, to amplify the signal from the photomultiplier, and for spectrometry, to determine the amplitudes of the pulses from the amplifier. Fig. 3.14 is a block diagram of the instrumentation used for scin-

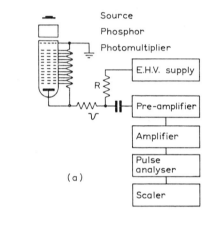

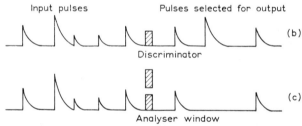

Fig 3.14 Block diagram of the instrumentation for a scintillation spectrometer (*a*), showing the use of the pulse analyser as a discriminator (*b*), and for the selection of pulses of defined amplitudes (*c*)

tillation spectrometry. The e.h.v. unit provides a stabilized high voltage to the dynode resistor chain of the photomultiplier. A high-quality amplifier, whose gain is stable and unaffected by pulse amplitude, is necessary for pulse amplification.

A preamplifier is positioned as close as possible to the photomultiplier. Its purpose is to permit the detector to be sited remotely from the amplifier and other electronic instruments and to effect impedance matching to the amplifier input. Amplitudes from the amplifier are measured using a pulse analyser. This instrument may be used as a discriminator, i.e. to reject all pulses whose amplitudes

Detection of Nuclear Radiation

lie below a selected level. Alternatively, it may be used to select only those pulses whose amplitudes lie between pre-selected levels. These functions are illustrated in Fig. 3.14. Finally, pulses selected by the pulse analyser are counted with a scaler.

The rates at which pulses of different amplitudes occur in the detector may be investigated in this way; a plot of counting rate against pulse amplitude is known as a *pulse spectrum*. In the case of the interaction of γ-rays in a sodium iodide crystal, a pulse spectrum such as that shown in Fig. E.14.1 (page 168) is obtained. This is generally called a *γ-ray spectrum*.

FURTHER READING FOR CHAPTERS 1-3

HARVEY, B. G.   *Nuclear Chemistry* (New Jersey, Prentice-Hall, 1965)
CHOPPIN, G. R.   *Nuclei and Radioactivity* (New York, Benjamin, 1964)
JENKINS, E. N.   *An Introduction to Radioactivity* (London, Butterworths, 1964)
FRIEDLANDER, G., KENNEDY, J. W. and MILLER, J. M.   *Nuclear and Radiochemistry*, 2nd ed. (New York, Wiley, 1964)
RUSK, R. D.   *Introduction to Atomic and Nuclear Physics* (London, Illiffe, 1965)
KALPAN, I.   *Nuclear Physics*, 2nd ed. (Reading, Mass., Addison-Wesley, 1963)
SEABORG, G. T.   *Man-made Transuranic Elements* (New Jersey, Prentice-Hall, 1963)
SWARTZ, C. E.   *The Fundamental Particles* (Reading, Mass., Addison-Wesley, 1965)
WILSON, B. J., Ed.   *The Radiochemical Manual* 2nd ed., (Radiochemical Centre, Amersham, 1966)

Chapter 4
The biological basis of radiation protection

It is widely known that ionizing radiation may be harmful to living organisms. This has, at times, been exaggerated by press reports so that the word "radioactivity" has achieved considerable notoriety as something that must be avoided at all costs. This is almost impossible to achieve, for Man has evolved under conditions of continuous exposure to low levels of ionizing radiation. Such exposures result from the naturally occurring isotopes carbon-14 and potassium-40 present in his body and uranium-238 and radium-226 present in the rocks and buildings that surround him. Cosmic radiation coming mainly from the sun also plays a significant contribution to this *background* radiation. The avoidance of, or protection from, such low doses of radiation is therefore a purely academic question.

With the advent of large-scale production of artificial radioisotopes, the radiation dose rate to the personnel involved in handling them and indirectly to the population at large has obviously increased. Strict protective measures have therefore been implemented so that this added dose to the individual above background levels is extremely small. It may well be argued that any dose above natural levels, however small, may be potentially harmful and should not be present. It must be appreciated, however, that the use of radioisotopes has been extremely beneficial to mankind in fields such as agricultural and medical research and the development of nuclear power. The very small possibility of harm from the resulting radiation exposure is more than compensated for by the benefits which have accrued. Another point to consider is that we fully appreciate the dangers of ionizing radiations and have, for many years, taken very strict measures to control exposure to them. This cannot be said for such hazards to human health as air and water pollution and bacterial immunity to antibiotics. Whereas we have been aware of these problems for some time, it is only recently that we have become concerned.

Radiation protection is almost as old as the discovery of ionizing radiations. It is concerned with the problems of reducing the harmful interaction of radiation with living organisms. Before discussing in

THE BIOLOGICAL CELL

The cell is the fundamental unit of all living organisms. Many millions of cells are grouped into separate working systems that form the human body. Their size and shape may be very varied. Fig. 1.1 illustrates the general features of a typical animal cell.

The *nucleus* is the 'brain' of the cell and controls much of the cell's chemistry or *metabolism*. It is spherical or ovoid in shape

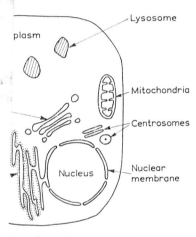

Fig. 1.1 Diagram of a typical animal cell

The endoplasm... is more extensive than shown in the diagram: the dots lining it are the ribosomes

and is surrounded by the nuclear membrane. Viewed under the microscope, the nucleus is seen to be composed of a mass of material called *chromatin*. At certain times during the cell's life, the chromatin becomes arranged into regular sausage-shaped structures called *chromosomes*. The importance of these will become apparent later.

Surrounding the nucleus is the *cytoplasm*—a material which has the consistency of thin jelly and contains 70–80% water. It is a complex mixture of mineral salts, amino acids, proteins and distinct physical structures called *organelles* which carry out various specific cellular functions. *Mitochondria*, for example, generate the chemical energy necessary for the cell to function, and the *Golgi apparatus* is a rather diffuse structure which is instrumental in cellular secretion.

The *lysosomes* are responsible for the digestion of substances taken into the cell body and the cell itself, when its useful life is over. The *centrosomes* are two structures which initiate cell division.

The majority of cells in the body take part in a continual succession of growth, division and death. Under normal conditions, cells reach a definite size and then divide to produce two daughter cells. The information that dictates the many chemical reactions going on within the parent cells is contained in the chromosomes, and this information must be accurately transmitted to the daughters in order that they may continue to carry out exactly the same functions as the parent in the organ or tissue of the body. In discussing radiation damage to biological systems we must consider the ways in which the information thus transmitted from parent to daughter cells may be altered by the radiation. Such alteration can profoundly change the physiological function of these new cells. It is first necessary, however, to look more closely at this information system.

THE MOLECULAR BASIS OF CELL CONTROL

The cells of Man may be broadly divided into two groups—those comprising the body organs and tissues and referred to as *somatic* cells, and those concerned solely with reproduction and termed *genetic* or germ cells. The somatic cells contain in their nucleus 46

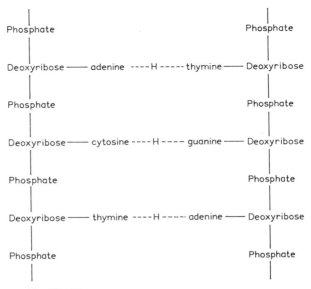

Fig 4.2 The structure of part of a molecule of DNA

Biological Basis of Radiation Protection 65

chromosomes arranged in pairs. The genetic cells possess 23 separate chromosomes. The reason for this difference will become apparent later, but let us first consider the molecular structure of these chromosomes, for it is these structures that possess the information for controlling the cell's function.

Chromosomes are made up of the protein, histone, and a nucleic acid in very close association. The nucleic acid is deoxyribonucleic acid (DNA); its general molecular structure is shown in Fig. 4.2, and its gross structure in Fig. 4.3. It is made up of two sugar-phosphate backbones in the form of a double helix. Along the backbones are attached four nucleotide bases—adenine (A), cytosine (C), guanine (G) and thymine (T). The linking of the two backbones is through hydrogen bonding and is illustrated in Fig. 4.4. It may be noted that, because of the chemical configuration of the bases, cytosine is always linked to guanine and adenine to thymine. It is the DNA molecule that carries the information necessary to control the manufacturing processes within the cell, which in turn determines its function.

The sites of manufacture of cellular materials are the small units called *ribosomes* situated in the cytoplasm (Fig. 4.1). To control cellular function, DNA itself does not leave the nucleus; instead the information is passed to another nucleic acid called messenger ribonucleic acid (m-RNA), which passes out through the nuclear membrane to the ribosomes. Here the information is translated and the manufacturing processes are initiated. The m-RNA is usually a single-stranded molecule and is very similar to a single backbone of the DNA molecule. It differs, however, in possessing the sugar ribose instead of the deoxyribose present in DNA. It has three nucleotide bases the same as the DNA molecule, but a fourth, uracil (U), replaces the thymine of DNA.

The m-RNA is manufactured from basic constituents in the cell and is built upon the DNA molecule, which acts as a template. Because of the strict relationship between adenine and thymine (or uracil in RNA), and guanine and cytosine, the m-RNA molecule must reflect the base sequence of the DNA molecule (Fig. 4.5(a)) This is the clue as to how the DNA molecule possesses information— it is related to the base sequence along its backbone. The sequence of the bases on both the DNA and RNA molecules forms a code of three-letter words. The four bases represent a four-letter alphabet, and each three-letter word, or triplet, is in general a specific code (*codon*) for an amino acid. A code using four letters can make a total of 64 triplets ($4 \times 4 \times 4$). As there are only 20 amino acids in the living system, there is a certain amount of duplication (the code is "degenerate"). A few of the triplets are, however, not equivalent to an amino acid but signify "end of message" as would be necessary

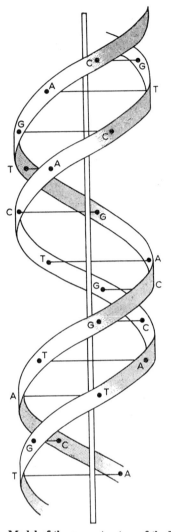

Fig 4.3 Model of the gross structure of the DNA molecule

The ribbons represent the sugar-phosphate backbone
and the letters represent the nucleotide bases
(see text). The central rod is an attempt to
give a clearer picture of the helix

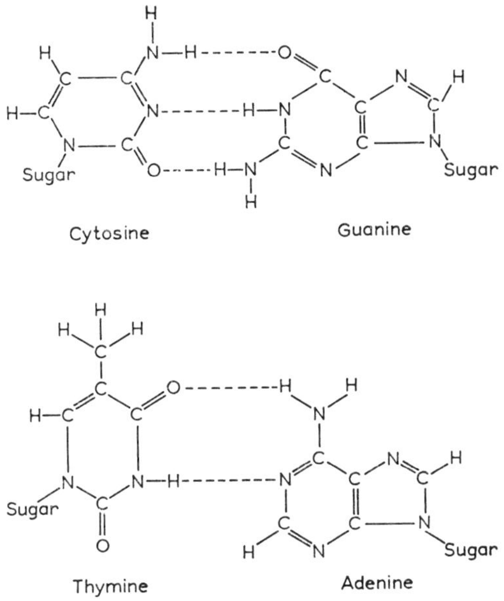

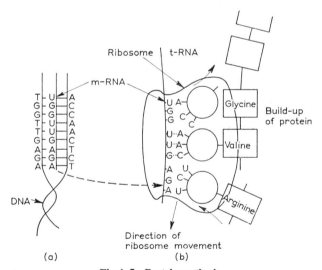

Fig 4.4 **Hydrogen bonding between nucleotide bases**

Fig 4.5 **Protein synthesis**

(a) The base sequence of m-RNA is dictated by the DNA molecule, which acts as template. The m-RNA migrates from the nucleus to the ribosomes.
(b) Amino acids with their respective t-RNA molecules arrive at the ribosome. The ribosome moves over the m-RNA, "scanning" small sections of the codon at a time and assembles those amino acids relevant to their t-RNA code into proteins.

with any coding system. A few examples of the codes and their equivalent amino acids are given in Table 6.

The coded information on the m-RNA now arrives at the ribosome. Here the code is translated and a series of amino acids are linked together to make a complex molecule called *protein*. The individual amino acids are brought to the ribosome to be incorporated

Table 6 Examples of the coding system on m-RNA

m-RNA code words	Amino acid
CGC or AGA	Arginine
GAU	Aspartic acid
GGU or GGA	Glycine
GUU	Valine
GAA	Glutamic acid
UGA	"End of message"

into the protein by another form of RNA called transfer-RNA (t-RNA). This molecule is a small helix, one end of which carries a triplet of nucleotide bases, and the other end, the amino acid corresponding to the code of that triplet. The triplet of the t-RNA is matched to that on the m-RNA at the ribosome, and the amino acids are linked up in the order required by the m-RNA (Fig. 4.5(b)). The t-RNA then separates from the amino acid to link up eventually with further similar amino acids. The complete protein synthesis is illustrated in Fig. 4.5. The protein formed is called an *enzyme*, which catalyses further complex chemical reactions in the cell. Many enzymes are produced in the cell, and each one is in general specific; i.e. it will catalyse only one particular chemical reaction. As an example, all animal tissues carry out a lactic acid fermentation in which one molecule of glucose is converted to two molecules of lactic acid. This is achieved in eleven chemical steps so requiring eleven specific enzymes. The total chemical activity of a cell is therefore extremely complex, and it is all controlled by the information coded on the DNA molecule in the nucleus through the agencies of messenger-RNA, transfer-RNA and the enzyme systems which are synthesized.

Any chemical change occurring in an enzyme, for example an amino acid not in the correct position, may render it useless in catalysing the particular reaction. Any change in the base sequence from the normal on the DNA molecule may mean that a particular amino acid is not produced. An example of such a condition is *sickle-cell anaemia*, a condition prevalent in Africa in which the red blood cells, instead of having a characteristic disc shape are in the form of a crescent, very much like a new moon. Haemoglobin is the

Biological Basis of Radiation Protection

material present in the red-blood cells which transports oxygen from lungs to tissues. Individuals suffering from sickle-cell anaemia possess a haemoglobin with very low efficiency in this vital oxygen transfer. The only difference between this haemoglobin and normal is that an amino acid, glutamic acid, in the normal is replaced by valine—a very small chemical change indeed, but with a very considerable physiological effect.

A change in the base sequence from the normal is called a *mutation*, and it has been estimated that 97% of all mutations are harmful. A mutation produced in a somatic cell will probably not be damaging to the organism as a whole, for this is only one cell out of the many millions constituting the tissue. The cell itself will probably not be able to carry out its correct function and may die or at least continue to live with reduced efficiency. If, however, a mutation occurs in a genetic cell—i.e. one that may be instrumental in producing a new individual—then this may be extremely hazardous indeed. Let us now discuss a few general concepts in the subject of genetics.

THE TRANSFER OF GENETIC INFORMATION IN CELL REPLICATION

We have said that, when a cell divides, the genetic information must be passed from the parent to the two daughter cells. It is therefore necessary to duplicate the information carried in the chromosomes of the parent cell. The important point about the DNA molecule is that it is self-replicating. During cell division, the two strands of the helix separate as shown in Fig. 4.6, and each strand acts as a template upon which a complementary strand is synthesized. Since a

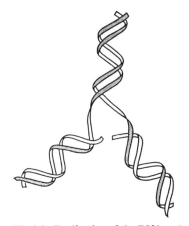

Fig 4.6 Replication of the DNA molecule

double helix of DNA can only be built up by hydrogen bonding between specified pairs of nucleotide bases, the result of the synthesis must be the production of a copy of the original molecule. All the chromosomes of the cell are copied in this way, and on cell division two sets of identical chromosomes are available for the daughter cells.

It was previously mentioned that the somatic cells of Man contain 23 pairs of chromosomes, one chromosome of each pair being derived from each parent. This, therefore, is the reason why genetic cells (*gametes*) have only 23 chromosomes. At fertilization, i.e. the fusion of male and female gametes, the 23 chromosome pairs are formed and the genetic characters (*genes*) of both parents are brought together. The total genetic make-up of the new individual is termed the *genotype*, and the characteristics that manifest themselves, the *phenotype*. When, for example, the new individual carries both the genes for brown and blue eye colour from its parents, it is the brown eye colour that makes its appearance, because the gene for brown eye colour is *dominant* over that for blue. This latter gene is known as *recessive*. The phenotype for blue eyes will arise only when the two recessive genes come together. When we speak of the gene for eye colour, we mean that series of bases on the DNA molecule that will dictate the enzymes that will catalyse the chemical reactions producing the pigments for eye colour.

MUTATIONS

At fertilization, a single cell is produced that will eventually give rise to the many millions making up the new individual. Any mutation in the information system of either gamete before this critical stage will be amplified and genetic abnormality will result. This may not be lethal to the individual, but will almost certainly have some detrimental effect on his life.

Point mutations, generally not associated with observable chromosome changes, result in a dominant gene changing to its recessive form. Such a *recessive mutation* will remain "hidden" until two parents carrying the same recessive gene have children. There will then be a distinct possibility that the mutation will become apparent. Fig. 4.7 shows diagrammatically how this may come about. Examples of recessive mutations are albinism, deaf-mutism and phenylketuria (metabolic disturbance resulting in mental retardation).

Dominant mutations are generally the result of chain breakage (Fig. 4.8) in the DNA molecule and may be observed in cells carrying them as structural damage to the chromosomes. Such mutations

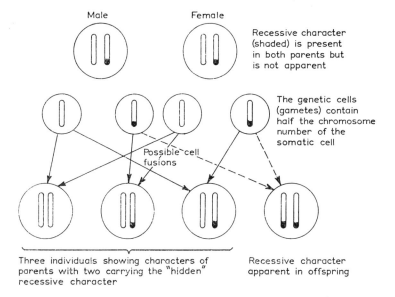

Fig 4.7 Appearance of a recessive mutation in the offspring

For simplicity only one pair of chromosomes has been considered

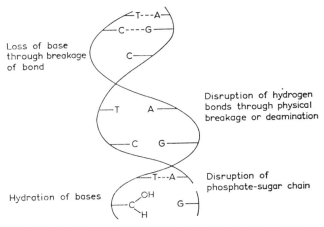

Fig 4.8 Structural damage to the DNA molecule by ionizing radiation

Bases signified by A (adenine), C (cytosine), G (guamine) and T (thymine)

will become apparent immediately in the offspring; examples are achondroplasia (dwarfness), dyslexia (word blindness) and migraine.

The more serious dominant mutations will tend to die out of a population rapidly since they diminish the individual's ability to survive. With recessive mutations, however, the defects may build up to some considerable extent in a population before they become observable.

THE EFFECTS OF IONIZING RADIATIONS ON BIOLOGICAL SYSTEMS

As ionizing radiations pass through living tissue, they lose energy until they eventually stop. As regards energy input and output, cells are in a very delicate state of balance, so the deposition of extra amounts of energy may seriously upset this balance. Although the whole chemistry of the cell is the target for ionizing radiations, it is the disruption of the large complex molecules such as DNA that will have the more immediate damaging effect on the cell. This damage is more efficiently brought about by an indirect reaction in which the radiation first interacts with the cell water, producing highly reactive oxidizing radicals. Some of the more important chemical reactions involved may be summarized as follows:

$$H_2O + \text{radiation} \rightarrow H_2O^+ + e^-$$
$$\rightarrow H\cdot + \cdot OH$$
$$H_2O^+ + H_2O \rightarrow H_3O^+ + \cdot OH$$
$$\cdot OH + \cdot OH \rightarrow H_2O_2$$

dissolved $O_2 + H\cdot \rightarrow HO_2\cdot$ (the perhydroxyl radical)

$$HO_2\cdot + HO_2\cdot \rightarrow H_2O_2 + O_2$$

All these products can irreversibly oxidize the DNA molecule and its bases, completely disrupting its organized structure (Fig. 4.8). Radiation may also directly break bonds in the DNA chain and cause rearrangement of its nucleotide bases, but such effects are only likely at very high radiation doses.

Whatever the particular effects of radiation on the DNA molecule, the general effect is the disruption of the information system. This may result in subtle changes in perhaps the characteristics of cell growth—the production of cancer cells—or simply cell death. It all depends very much on the dose the cells have received. If a mutation is induced in a germ cell that ultimately gives rise to a new individual, then the injury can range from such harmless abnormalities as inability to taste the chemical phenylthiocarbamide to gross abnormalities of the brain.

Biological Basis of Radiation Protection

In considering the genetic effects of radiation, it is important to appreciate that genes may mutate spontaneously. Radiation is an agent, along with others such as nitrogen mustards* that will tend to increase this natural mutation rate. Such agents will not produce new mutations that have not been observed naturally, they will just tend to increase their frequency.

This then is the basic hazard of ionizing radiation. At certain levels it can induce mutations and cancers such as leukaemia, and can shorten life expectancy. At very high doses (accidental levels) it can produce severe damage to whole organ systems of the body. Radiation, however, is also of great benefit to mankind and it is therefore necessary to balance these risks against such benefits.

Occupationally exposed radiation workers must anticipate a certain risk attending their work, something that is in common with many other occupations. This risk, however, must be acceptable to the individual. The extent of radiation exposures which may be regarded as acceptable risks, called *maximum permissible levels* (MPLs) have been agreed by the International Commission on Radiological Protection [1, 2]. In the next chapter we shall deal with the broad aspects of these recommendations for the limitation of radiation dose.

* $RN(CH_2 CH_2 Cl)_2$

REFERENCES

1 *Recommendations of the International Commission on Radiological Protection* (ICRP Publication 2, Pergamon 1959)

2 *Recommendation of the International Commission on Radiological Protection* (ICRP Publication 9, Pergamon 1965)

General reading
Asimov, I.   *The Genetic Code* (John Murray 1964)
Butler, J. A. V.   *The Life of the Cell* (Allen & Unwin 1964)

Special reading
Purdom, C. E.   *Genetic Effects of Radiations* (Newnes 1963)
Dubinin, N. P.   *Problems of Radiation Genetics* (Oliver & Boyd 1964)

Chapter 5

Radiation dosimetry and maximum permissible levels of dose

Radiation protection poses two basic questions. Firstly, how does one determine what level of radiation exposure may be classed as safe, and secondly, how does one limit exposure to this level when working with radioactivity? The first question is answered by the work of the International Commission on Radiological Protection (ICRP) and will be discussed as the main topic of this chapter. The second question involves common laboratory practice and will be discussed in detail in Chapter 6. We shall consider first the units employed for radiation protection.

DEFINITIONS OF RADIATION QUANTITIES AND UNITS [1]

In Chapter 3 we saw that, when radiation passes through matter, it causes ionization and so loses energy. This energy loss generally results in a temperature rise in the absorber material. The concept of *dose* is used to define the extent to which matter has been exposed to radiation, and it should therefore be possible to measure this dose in terms of ionization, energy uptake or temperature rise in the absorber. In practice it is easier to measure the extent of ionization which occurs in air than to measure the very small temperature increase which results when materials are exposed to radiation. The unit of radiation exposure is therefore defined in terms of the amount of ionization that is produced in air. If a radioactive source which emits γ-rays is placed in air then it will cause ionization in the air surrounding it. If we measure the amount of electrical charge produced in a known mass of air then we have a method of measuring *exposure dose*. The formal definition of this dose is as follows:

"*Exposure* (X) is the quotient of ΔQ by Δm, where ΔQ is the sum of the electrical charges on all the ions of one sign produced in air, when all the electrons liberated by photons in a volume element of air whose mass is Δm are completely stopped in air":

$$X = \frac{\Delta Q}{\Delta m} \qquad (5.1)$$

Dosimetry and Maximum Dose

The unit of exposure is called the *röntgen* (R):

$$1\text{ R} = 2 \cdot 58 \times 10^{-4} \text{ C/kg}$$

This definition of exposure is illustrated in Fig. 5.1. The volume element of air of mass Δm is represented by a cube which is exposed to a source of photons (X- or γ-radiation). A photon which interacts with an atom in this sensitive volume will produce secondary electrons as discussed in Chapter 3. These electrons wander through the cube

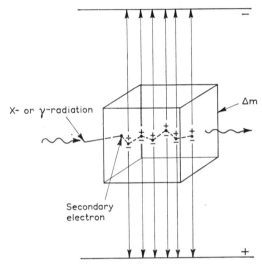

Fig 5.1 Ionization in a sensitive volume of air

of air producing further ion pairs. If a potential difference is applied between electrodes placed in the sensitive volume then the positive ions will be collected at the cathode and the negative ions at the anode. The definition of exposure is concerned only with the sum of the charges of either sign, not both.

The röntgen has one major disadvantage as a dose unit in that it applies only to X- and γ-radiation absorbed in air. This has serious limitations in radiological protection where other radiations may be involved and the dose is to biological tissue instead of air. Because of these difficulties, the concept of *absorbed dose* was introduced. Its definition is as follows:

"Absorbed dose (D) is the quotient of ΔE_D by Δm, where ΔE_D is the energy imparted by ionizing radiation to the matter in a volume element of mass Δm":

$$D = \frac{\Delta E_D}{\Delta m} \tag{5.2}$$

The unit is the *rad* (*R*adiation *A*bsorbed *D*ose):

$$1 \text{ rad} = 100 \text{ ergs/g} = 10^{-2} \text{ J/kg}$$

This unit covers all kinds of radiation which may be absorbed in all types of material. It is really equivalent to the heat appearing in the material, and because of this it is extremely difficult to measure directly. When 1 g of soft tissue is exposed to 1 R of γ-rays, approximately $9 \cdot 5 \times 10^{-6}$ J of energy are deposited. Because this value is so close to the 10^{-5} J deposited per gramme when 1 rad is absorbed, it may be assumed that, for soft tissue, exposure to 1 R of γ-rays is approximately equal to an absorbed dose of 1 rad.

Table 7 The linear energy transfer and quality factor for various radiations in water

Taken from the Recommendations of the International Commission on Radiological Protection, *I.C.R.P. Publication 9*, by courtesy of Pergamon Press.

Type of radiation	L.E.T., keV/μm	Q.F.
β- or γ-radiation	3·5 or less	1
	7·0	2
Slow neutrons	23	5
Fast neutrons or α-radiation	53	10
Heavy recoil particles	175	20

In radiation protection the concept of absorbed dose is still not completely satisfactory, for heavily ionizing radiations such as α-particles are more effective in producing cell damage than γ-rays or electrons. This is because an α-particle emitted within a cell will deposit its energy entirely within the cell, whereas γ-rays are little attenuated by any biological tissue. The greater the amount of energy deposited per cell the greater is the biological damage. The effectiveness of the radiation in producing biological damage is measured in terms of its *quality factor* (Q.F.). This factor is used only for radiation protection purposes and is solely dependent on the quantity known as the *linear energy transfer* (L.E.T.) of the radiation, which is defined as the energy lost per unit path length of radiation, and has a special unit, the kilo-electronvolt per micrometre. In general, the higher the L.E.T., the greater will be the effectiveness of the radiation for producing biological damage, and consequently the quality factor increases with increasing L.E.T. The relationship between L.E.T. and Q.F. is shown in Table 7. Quality factor has no units. Reference to the table shows that fast neutrons

Dosimetry and Maximum Dose

are 10 times more effective in producing biological damage than is γ-radiation.

It is now possible to define a unit of dose which will take into account the biological effectiveness of the various kinds of radiation. This is necessary because in radiation protection every kind of radiation may be encountered. To arrive at this final unit, the absorbed dose (D) is multiplied by the quality factor. This results in a term called the *dose equivalent* (*D.E.*):

$$D.E. = D \text{ (rad)} \times Q.F. \tag{5.3}$$

The unit of dose equivalent is the *rem* (*R*öntgen *E*quivalent *M*an). To make the concept clearer, let us assume that the rad is used as the unit of dose for radiation protection. In such a case it would require a dose of 50 rads due to γ-rays to produce the same damage as a dose of 5 rads due to fast neutrons. A mixed flux of neutrons and γ-rays would thus present complications in expressing a useful total radiation dose. By using the rem as the unit of dose, however there is no complication: for example, 5 rems of γ-rays will produce the same biological damage as 5 rems of neutrons. In expressing maximum permissible dose levels (M.P.L.s) for radiation protection purposes, therefore, the only units used are the rem and sub-multiples such as the millirem (mrem).

ABSORBED DOSE FROM RADIOISOTOPES WITHIN THE BODY

Exposure to ionizing radiations can come from two sources: those from outside the body constituting an *external* hazard and those inside the body constituting an *internal* hazard. External hazards are relatively straightforward in their assessment, since radiation doses from external sources can easily be measured or calculated (see Chapter 6). Internal hazards, however, present considerable problems in their assessment, for when radioisotopes are ingested they are usually distributed amongst one or more organs of the body and contribute a much higher dose to these organs than to the rest of the body. The most likely means of ingesting radioisotopes are

1. Inhalation of contaminated air
2. Drinking contaminated water
3. Entry through open wounds

The particular organs that concentrate the ingested isotope are called *critical organs*. The criteria employed to classify an organ as critical for a particular radioisotope are

1. The fraction of the total body content of the radioisotope which accumulates in the organ
2. The essentiality of the organ for the proper functioning of the body
3. The sensitivity of the organ to damage by radiation
4. The route of entry into the body and the possible damage caused en route

The actual absorbed dose received by the critical organ will depend upon the emissions of the radioisotope present and the size and shape of the organ. Because of the variable size of these organs in a population, the ICRP have adopted a "standard man" for whom the average masses and dimensions of all the organs are listed [2].

Radioisotopes present in the body are eliminated by biological processes in such a way that the quantity present in the body decreases exponentially with time. They possess a biological half-life (T_b), which is the time taken to eliminate half the amount of isotope from the body by biological processes. The overall elimination of the isotope will be a combination of the rates of biological elimination and radioactive decay and is represented by the effective half-life (T_{eff}), defined by

$$\frac{1}{T_{eff}} = \frac{1}{T_b} + \frac{1}{T_{1/2}} \tag{5.4}$$

in which $T_{1/2}$ is the radioactive half-life of the isotope.

If a radioisotope is ingested at a constant rate, for example by an occupational worker who is continuously breathing slightly contaminated air whilst at work, then eventually an equilibrium will be established when the rate of elimination of the isotope from his body becomes equal to the rate of ingestion. The amount of isotope in the body then becomes constant. The quantity of radioisotope present in the body at any one time is known as the *body burden*.

MAXIMUM PERMISSIBLE LEVELS OF RADIATION DOSE

The International Commission on Radiological Protection (ICRP), established in 1928, is an organization whose policy is to prepare and continually modify recommendations on the basic principles of radiation protection. Their efforts have culminated in an ever-growing series of reports [2-5] whose aim is to serve as a guide, not only to those who are actively engaged in radiation work, but also

Dosimetry and Maximum Dose

to those government departments whose concern is the protection of members of the whole population. In the United Kingdom these government departments include the Department of Education and Science, whose concern is the control of radiation exposure to pupils and students, and the Medical Research Council, who make recommendations for the protection of both occupationally exposed workers and the population at large. A new National Radiological Protection Board has been established under the Radiological Protection Act, 1970. The function of the Board will be to advise on matters of radiological protection.

The ICRP have formulated the concept of tolerable or permissible levels of radiation exposure. Since the objective of radiation protection is "to prevent or minimize somatic injuries and minimize the deterioration of the genetic constitution of the population", many of their publications are concerned with the establishment of *maximum permissible levels* (M.P.L.s).

The recommended maximum permissible accumulated dose for occupationally exposed radiation workers is given by

$$D = 5(N - 18) \text{ rems} \tag{5.5}$$

where D is the tissue dose and N is the age of the exposed worker in years. This formula applies to the gonads and blood-forming organs, for these are the most radiosensitive organs in the body. The formula shows that a radiation worker aged 30 years could have accumulated a maximum dose of 60 rems. If he had started this work at the age of 18 years, he could have accumulated the dose at a maximum rate of 5 rems every year. This level of dose forms the basis of maximum levels employed in industrial isotope laboratories; a year of 50 working weeks is equivalent to an average exposure of 100 mrems/week. To prevent the accumulation of 5 rems in one exposure, the ICRP impose a limiting dose rate of 3 rems in 13 consecutive weeks. No further limit is imposed, so the worker could be subjected to 3 rems at a single exposure, so long as no other radiation exposure occurred for the following 13 weeks. It is important to realize that these M.P.L.s do not include exposures to background radiation or medical procedures.

From eqn. (5.5), a radiation worker at an assumed retirement age of 65 years could have accumulated a dose of 235 rems. This dose, provided that it is received at a rate of not more than 3 rems/13 weeks, is considered to have a negligible probability of causing somatic or genetic injury to the worker or his children.

The maximum permissible dose of 5 rems/year applies only to the most radiosensitive organs; other areas of the body that are more radioresistant or contain no vital organs may be exposed to the somewhat higher doses listed in Table 8. The values quoted in the table apply to occupational workers throughout most of their lifetime. As regards the genetic effects of radiation, it is the first 30 years of life that constitute the greatest hazard. With this in mind, the ICRP have recommended that the average dose received by a whole population, other than occupationally exposed workers (who

Table 8 Maximum permitted doses for various body organs

Taken from The Recommendations of the International Commission on Radiological Protection, *I.C.R.P. Publication 9*, by courtesy of Pergamon Press.

Organ	Maximum permitted doses for occupationally exposed workers (*per year and per 13 consecutive weeks*)	Maximum permitted doses for occupationally exposed workers (*based on a 50-week year*)
Gonads and red bone marrow	5 rems/year 3 rems/13 weeks	100 mrems/week
Skin, bone thyroid	30 rems/year 8 rems/13 weeks	600 mrems/week
Hands, forearms, feet, ankles	75 rems/year 20 rems/13 weeks	1500 mrems/week
Other single organs	15 rems/year 4 rems/13 weeks	300 mrems/week

would not normally exceed 0·2% of the population), should not exceed 2 rems up to the age of 30 years, which is considered to be the mean age of child bearing. This genetic dose for the public at large means that, if every member was exposed to 2 rems up to the age of 30, the genetic burden imposed upon the population would be considered tolerable when compared to the benefits derived from the uses of radiation.

MAXIMUM PERMISSIBLE BODY BURDENS FOR ISOTOPES DEPOSITED IN THE BODY

For those workers who only handle sources that present an external hazard, it is sufficient to consider only the M.P.L.s quoted in Table 8. For workers, however, who handle radioactive sources that present not only an external but also an internal hazard due to their possible ingestion, these M.P.L.s must be modified to take account of any dose due to internally deposited isotopes. Because individual isotopes tend to concentrate in particular organs of the body, the ICRP quote *maximum permissible body burdens* for all radionuclides, some of which are given in Table 9 (page 82). These are the quantities of the isotope which when present in the body give the maximum permissible dose of 300 mrems/week (see Table 8) to the critical organ. Also quoted in Table 9 are *maximum permissible concentrations* of radioisotopes in air and water (MPC_a and MPC_w). These values are quoted for 40-hour and 168-hour weeks. The "standard man" breathing or drinking air or water so contaminated over such periods in each week would eventually accumulate a maximum permissible body burden. It is again important to note that the quoted body burdens assume that there is no contribution from external radiation. With weak β-emitters such as carbon-14 and sulphur-35, this is so, but for β/γ-emitters such as iodine-131 there is both an internal and an external radiation hazard.

TOXICITY CLASSIFICATION OF RADIOISOTOPES

The International Atomic Energy Agency(I.A.E.A.)[6] have proposed a toxicity classification for radionuclides based on such factors as the identity of the critical organ, the effective half-life of the isotope and the quality factor of radiation emitted. The classification for some selected isotopes is given in Table 10 (page 84).

Class I are nuclides of very high toxicity when present in the body. A large number of them are α-emitters. The α-sources do not present an external hazard because of their very limited penetration in the skin. When taken into the body, however, all their energy is deposited within a few micrometres of their emission. Cells a few micrometres in diameter would suffer extensive damage from the deposition of such energies. Class II and III radionuclides are of medium toxicity and constitute the majority of the β/γ-emitting isotopes used as tracers. Class IV, the low toxicity nuclides, includes the naturally occurring radioisotopes and the radioactive isotope of hydrogen (tritium, or hydrogen-3).

This classification applies to the radioisotopes in the elementary or simple inorganic chemical forms, and is not necessarily valid for complex labelled molecules. Some of these complex molecules act as a passport for the element into the chemistry of the cell, and labelled molecules introduced in this manner may be responsible for a considerable amount of cellular damage. The molecule thymidine (thymine plus sugar) is incorporated into the base sequence of the DNA molecule. This molecule labelled with tritium could therefore inflict considerable damage when incorporated in a cell. Tritium-labelled water, however, would be responsible for less damage because of its general distribution throughout the body.

Table 9 Some data on the internal deposition of isotopes

Isotope	Half-life, $T_{1/2}$	Critical organ	Effective half-life, T_{eff}	Maximum permissible body burden, q (μCi)
Tritium (H-3)	12·26 y	Body tissue	12 d	1000
Carbon-14	5760 y	Fat	40 d	300
Sodium-22	2·6 y	Total body	11 d	10
Sodium-24	15·0 h	Total body	14·4 h	7
Phosphorus-32	14·3 d	Bone	13·5 d	6
Sulphur-35	87·2 d	Testis	76·4 d	90
Potassium-42	12·4 h	Total body	12·9 h	10
Calcium-45	165 d	Bone	162 d	30
Iron-59	45 d	Spleen	42 d	20
Cobalt-60	5·25 y	Total body	9·5 d	10
Zinc-65	245 d	Total body	13 d	60
Strontium-89	50·5 d	Bone	50·4 d	4
Strontium-90	28 y	Bone	17 y	2
Iodine-131	8 d	Thyroid	7·6 d	0·7
Caesium-137	30 y	Total body	70 d	30
Barium-140	12·8 d	Bone	10·7 d	4
Cerium-144	285 d	Bone	243 d	5
Lead-210 (Radium-D)	21 y	Kidney	1·9 y	0·4
Lead-212 (Thorium-B)	10·6 h	Kidney	10·6 y	0·02
Polonium-210	138·4 d	Spleen	42 d	0·03
Radium-226	1620 y	Bone	44 y	0·1
*Natural thorium	—	Bone	—	0·01 = 50 mg thorium
*Natural uranium	—	Kidney	—	0·005 = 16 mg uranium
Plutonium-239	24 000 y	Bone	175 y	0·04

* Chemical toxicity has not been taken into account in the above table except for natural thorium and uranium, where the chemical toxicity rather than the absorbed dose limits the body burden, q.
  Data from Recommendations of the International Commission on Radiological Protection Report of Committee II on permissible dose for internal radiation, *I.C.R.P. publication* 2, by courtesy of Pergamon Press.

Table 9 (Continued)

Maximum permissible concentration in air for exposure 40 h each week, MPC_a ($\mu Ci/cm^3$)	Maximum permissible concentration in water for exposure 40 h each week, MPC_w ($\mu Ci/cm^3$)	Maximum permissible concentration in air for exposure 168 h each week, MPC_a ($\mu Ci/cm^3$)	Maximum permissible concentration in water for exposure 168 h each week, MPC_w ($\mu Ci/cm^3$)
5×10^{-6}	0.1	2×10^{-6}	0.03
4×10^{-6}	0.02	10^{-6}	8×10^{-3}
2×10^{-7}	10^{-3}	6×10^{-8}	4×10^{-4}
2×10^{-6}	0.01	6×10^{-7}	4×10^{-3}
7×10^{-8}	5×10^{-4}	2×10^{-8}	2×10^{-4}
3×10^{-7}	2×10^{-3}	9×10^{-8}	6×10^{-4}
3×10^{-6}	0.02	10^{-6}	8×10^{-3}
3×10^{-8}	3×10^{-4}	10^{-8}	9×10^{-5}
10^{-7}	4×10^{-3}	5×10^{-8}	10^{-3}
4×10^{-7}	4×10^{-3}	10^{-7}	10^{-3}
10^{-7}	3×10^{-3}	4×10^{-8}	10^{-3}
3×10^{-8}	3×10^{-4}	10^{-8}	10^{-4}
3×10^{-10}	4×10^{-6}	10^{-10}	10^{-6}
9×10^{-9}	6×10^{-5}	3×10^{-9}	2×10^{-5}
6×10^{-8}	4×10^{-4}	2×10^{-8}	2×10^{-4}
10^{-7}	6×10^{-3}	4×10^{-8}	2×10^{-3}
10^{-8}	0.2	3×10^{-9}	0.08
10^{-10}	4×10^{-6}	4×10^{-11}	10^{-6}
2×10^{-8}	6×10^{-4}	6×10^{-9}	2×10^{-4}
5×10^{-10}	2×10^{-5}	2×10^{-10}	7×10^{-6}
3×10^{-11}	4×10^{-7}	10^{-11}	10^{-7}
2×10^{-12}	3×10^{-5}	6×10^{-13}	10^{-5}
7×10^{-11}	2×10^{-3}	3×10^{-11}	6×10^{-4}
2×10^{-12}	10^{-4}	6×10^{-13}	5×10^{-5}

Table 10 Classification of radionuclides according to toxicity (an abbreviated table)

Class I *High Toxicity*
Strontium-90, Polonium-210, Radium-226, Plutonium-239

Class II *Medium Toxicity—Upper sub-group A*
Sodium-22, Calcium-45, Cobalt-60, Strontium-89, Iodine-131, Caesium-137, Barium-140, Cerium-144, Lead-212

Class III *Medium Toxicity—Lower sub-group B*
Carbon-14, Sodium-24, Phosphorus-32, Sulphur-35, Potassium-42, Iron-59, Zinc-65

Class IV *Low Toxicity*
Tritium (Hydrogen-3), Natural Thorium, Natural Uranium

REFERENCES

1 *Radiation Quantities and Units*, International Commission on Radiological Units and Measurements (ICRU) Report 10a, Handbook 84 (National Bureau of Standards, 1962)

2 Recommendations of the International Commission on Radiological Protection, Publication 2, *Report of Committee II on Permissible Dose for Internal Protection* (Pergamon 1959)

3 Recommendations of the International Commission on Radiological Protection, Publication 6, amended 1959, revised 1962 (Pergamon 1964)

4 Recommendations of the International Commission on Radiological Protection, Publication 9, adopted 1965 (Pergamon 1966)

5 Recommendations of the International Commission on Radiological Protection, Publication 10, *Report of Committee IV on Evaluation of Radiation Doses to Body Tissues from Internal Contamination due to Occupational Exposure* (Pergamon 1968)

6 Technical Reports Series No. 15 (Vienna, IAEA, 1963)

General reading

REES, D. J. *Health Physics* (Butterworth 1967)

Chapter 6
Practical methods of radiation protection

At the beginning of Chapter 5 we posed two questions on radiation protection. We have now answered the first question, which was concerned with the establishment of safe levels of radiation exposure. We now consider the second question: how may we carry out work with radioactive materials and keep within the appropriate recommendations for maximum levels of radiation exposure? It is necessary to consider both the internal and external hazards which arise in the use of a given radioactive substance. The internal hazard arises from ingestion of the radioactive material and is controlled by the prevention of such ingestion. This is accomplished by the techniques adopted in laboratory practice, and these are discussed later in this chapter. Regarding protection from external hazards, it is first necessary to assess the extent to which the radiation source presents a dose hazard. This can be determined either by measuring the dose rate from the source with an instrument called a *dose-rate meter* or by calculating the dose rate. In many cases and certainly for the source activities which are used in the experiments described later, it is sufficiently accurate to estimate dose rates by calculation. Since α-particles are absorbed by a few centimetres of air, α-particle sources do not present an external hazard. The methods for calculating the dose rates due to β/γ-emitting isotopes are described below.

CALCULATION OF EXPOSURE DOSE-RATES

For γ-ray emitting sources

The exposure dose-rate (X) in milliröntgens per hour (mR/h) at a distance d centimetres from a γ-ray source of activity A microcuries (μCi) is given by the equation

$$X = \frac{A \times \Gamma}{d^2} \text{ milliröntgens/hour} \qquad (6.1)$$

where Γ is a constant called the *specific γ-ray constant*. The units in which Γ is expressed must be compatible with those used for exposure (X), activity (A), and distance (d) in eqn. (6.1), i.e. it must be equal to the dose rate in milliröntgens per hour for a 1 μCi source at a distance of 1 cm. Values of this constant for various isotopes are given in Appendix 5.

For applications to sources of higher activity it is useful to express Γ in terms of röntgens and millicuries rather than milliröntgens and microcuries, which makes no difference to the numerical value.

The following example of the calculation of the dose-rate due to a 100 μCi cobalt-60 source at a distance of 50 cm will make the application of eqn. (6.1) and the use of the specific γ-ray constant clear.

Since Γ for cobalt-60 = 13·2 mR μCi^{-1}h^{-1}cm^2, then

$$X = \frac{100 \times 13\cdot2}{50^2} = 0\cdot52 \text{ mR/h}$$

Since Γ for cobalt-60 = 13·2 R mCi^{-1}h^{-1}cm^2, then

$$X = \frac{0\cdot1 \times 13\cdot2}{50^2} = 0\cdot000\ 52 \text{ R/h} = 0.52 \text{ mR/h}$$

For β-emitting sources

Since β-particles undergo air absorption the amount of which is dependent upon their energy, it is not possible to derive an accurate expression for the exposure rate due to a β-particle source. The following approximate formula for the absorbed dose (D) in millirads/hour at a distance d centimetres from a β-particle source of activity A microcuries is, however, useful:

$$D = \frac{10^3 A}{3d^2} \text{ millirads/hour} \tag{6.2}$$

This expression is correct to within 10% over distances up to 30 cm for β-energies of 1·5 MeV upwards.

The calculation of the absorbed dose-rate from a 10 μCi source of phosphorus-32 at 5 cm is as follows:

$$D = \frac{10^4}{3 \times 5^2} = 133 \text{ mrad/h}$$

METHODS OF RADIATION PROTECTION FOR SOURCES PRESENTING AN EXTERNAL HAZARD

Under this heading, we have the problem of dealing with closed (or sealed) sources which present no contamination hazard but

Practical Methods of Radiation Protection

constitute a hazard simply from their radiation. The radiation dose-rate to an operator handling such sources may be reduced by the following methods.

The use of minimal amounts of activity

Most radiation detectors are extremely sensitive and it is unnecessary to use sources of greater activity than can readily be detected. Activities of a few microcuries are adequate for most of the experimental work described later.

The use of some form of shielding

By placing an absorber or shield between the source and the operator the radiation dose can be significantly reduced. Table 11 gives some

Table 11 Shielding required for different types of radiation

Type of radiation	*Shielding required*	*Remarks*
α-emitters	None	α-particles are stopped in centimetres of air or within a few micrometres of tissue and present no external hazard
β-emitters	Aluminium, glass, Perspex	7 mm thickness of these materials will completely absorb the β-intensity from phosphorus-32
γ-rays	Lead, concrete	The thickness of shield required to reduce γ-radiation from cobalt-60 to $\frac{1}{10}$ of its initial intensity is about 5 cm of lead or 30 cm of concrete
Thermal neutrons	Paraffin wax or water, cadmium, boron	

guidance on the shielding required for different radiations.

Maintaining a safe distance from the source

Protection from the γ-radiation due to sources of the millicurie level or above may require the use of shielding and can be very expensive. An alternative to the use of shielding is to increase the distance between the source and the operator. It may be noted that γ-radiation obeys the inverse-square law

$$I_d \propto \frac{1}{d^2}$$

where I_d is the intensity of the radiation at distance d from the source (see Experiment 12). Eqn. (6.1) incorporates this law and shows that the dose rate due to a γ-ray source is inversely proportional to the square of the distance from the source. The provision of adequate distance between the operator and the source, however, may involve the use of remote-handling devices.

Reducing time of exposure
Where both shielding and maintaining a safe distance are impracticable, work with an active source of radiation should be conducted in the shortest possible time. The only source used in the experiments described later to which this might apply is the neutron source used in Experiment 22.

METHODS OF PROTECTION FROM INTERNAL RADIATION HAZARDS

Some sources presenting internal hazards also contribute an external hazard from their associated radiations. With β/γ-emitting isotopes such as iodine-131 and energetic β-emitters such as phosphorus-32, the precautions discussed above must be borne in mind as well as the following, which are designed to minimize the risk from radioisotope ingestion.

1. *Use of material of minimum toxiticy.* By reference to Table 10 an isotope should be selected from as low a toxicity class as is campatible with the other requirements of the work being carried out.

2. *Use of small levels of activity*

3. *Use of containment.* All work with open sources must be conducted in a confined area such that in the event of a spillage the material is contained and cannot contaminate the rest of the working area.

4. *Cleanliness.* In all work with open sources of radioisotopes, cleanliness is vital to prevent contamination of apparatus, clothing and the person. The hands should always be washed on leaving a radioisotope laboratory.

Practical Methods of Radiation Protection

5. *Use of protective clothing.* There is always the possibility of contamination through accidental spillage, and it is essential not to contaminate the skin and clothing. In all work with open sources, therefore, laboratory coats and gloves should be worn.

LABORATORY PROCEDURES FOR HANDLING RADIOISOTOPES

Closed (sealed) sources

Small closed sources of the order of a few microcuries in activity may be handled with impunity, provided that tongs or tweezers are used so that actual contact with the sources and the consequent high dose rate to which the fingers would be exposed are avoided. Millicurie sources may be used only for limited periods without the use of shielding, and remote-handling tongs must be used for manipulating them. The use of higher-activity sources will in general require the provision of considerable shielding.

Open (unsealed) sources

The use of open radioisotope sources, i.e. radioactive materials in the form of solutions, gases or solid materials which are not contained, involves an internal as well as an external radiation exposure hazard. This is because of the possibility that they may contaminate water, air or surfaces, and consequently may be ingested or gain accidental entry into the body, for example through wounds or from splashes on the skin. The possible dangers arising from entry into the body are specific for each radioisotope. Naturally the external hazard involved is the same in magnitude as that from similar, closed sources.

For chemical and biological work, radioactive materials are generally used at microcurie levels of activity. Ultimately all the active material used will become active waste, and the amount of activity in the waste will depend upon the extent to which the radioisotopes present have decayed. At all stages in the experimental work, until the active waste has been safely disposed of, there is the possibility of the spread of contamination and hence of the entry of active materials into the body. The objectives of the procedures adopted in handling open sources of radioisotopes are to control and limit this spread of contamination, and to minimize the risk of accidental entry of active materials into the body during the manipulations.

In school work, open α-particle sources of high specific activity, because of their very great biological hazard when ingested, are not permitted. The only α-emitting radionuclides which may be used as open sources are the low-specific-activity natural materials uranium and thorium as chemical reagents.

Work with β/γ-emitting isotopes in microcurie amounts is generally possible without any very elaborate shielding. If the radioisotope involved forms volatile compounds, or if there is a risk of entrainment of radioactive material in vapours or gases produced by the chemical processes involved, then the manipulations must be carried out in a fume cupboard. The following notes on procedures apply to manipulations involving β/γ-emitters or uranium and thorium compounds.

Use of protective gloves

The use of rubber or disposable plastic gloves can help to reduce the danger of ingestion of active material through contamination of the hands. The gloves should be of the thin latex surgical type, with a slightly roughened surface to facilitate handling apparatus, and should be a comfortable fit. Rubber gloves should have a long life if they are dried thoroughly before storage. Because such gloves are to be used many times, they must be put on and taken off in such a way that the hands do not come into contact with the outside surfaces, which should always be regarded as contaminated. After being taken off, they should be allowed to dry on the inside surfaces before being inverted and stored. A dusting talc on the hands and on the inside of the gloves will absorb perspiration and increase comfort when wearing them. The use of disposable gloves obviates the great care needed when using rubber gloves and is in many respects far more satisfactory and economical. It must be remembered, though, that a careless glove technique, in which contamination may spread to the inside of the gloves where it can readily be absorbed on moist hands, can be more hazardous than not wearing gloves at all. Gloves should be worn for all operations involving isotopes in solution except those with small amounts of short-lived radioisotopes of low biological hazard. They should not be worn whilst operating counting equipment or for handling liquid counters, since this would increase the risk of contaminating such equipment.

Double containment

The possibility of spreading contamination as a result of spillages and breakages is greatly reduced by using double containment whenever practicable. All bottles containing radioisotope stock solutions should be placed in metal cans and the space between the bottle and the can packed with absorbent wadding.

The principle of double containment should be extended to general working by the use of large enamelled trays. Trays measuring about 60 cm × 45 cm with a 5 cm rim are suitable and should be lined

Practical Methods of Radiation Protection 91

with a sheet of polythene, on top of which is a layer of absorbent paper. The paper will absorb drops and spills of active liquids and yet prevent contamination of the enamel which may be difficult to remove. Only apparatus containing radioactive materials or contaminated articles should be placed inside the tray. All reagents and other necessary apparatus should be placed on the bench beside the tray. This segregation prevents contamination of clean articles in the event of a spill and also facilitates the subsequent decontamination of equipment. It is useful to place two large beakers in a corner of the tray, one for the collection of small amounts of active liquid waste and the other for paper tissues, filter papers, etc. Paper tissues are useful for mopping up spills of liquid, for wiping the outside of pipettes and for laying contaminated articles on. They may also be used for handling contaminated articles when gloves are not being worn. A typical working layout is shown in Plate 5.

Addition of carriers

Many stock solutions of radioisotopes have very high specific activities, often of the order of 10^5 Ci per gram of isotope. After dilution and dispensing of a few microcuries of such a solution for a tracer experiment, the mass of material used would probably be of the order of 10^{-10} g. Chemical manipulations with such small amounts of material are very difficult, and unwanted side effects may occur. For example, if a solution containing 10^{-10} g of active phosphate were to be transferred from one beaker to another, it is possible that most of the active material would remain in the first beaker, being absorbed upon the surface of the glass. It is therefore necessary to add an inactive carrier, generally an inactive isotope of the same element, to such solutions during dilution and dispensing. The addition of as little as 1 mg of inactive phosphate in the above example makes the ratio of inactive to active phosphorus atoms about $10^7:1$. Absorption of 10^{-10} g of phosphate on glassware would then represent a negligible loss of active material. In the preparation of small sources by evaporation of carrier-free solutions, about 0·1 mg of carrier should be added to prevent losses by entrainment in the vapour during evaporation. Solutions of inactive carriers may also be used for decontamination of apparatus, exchange taking place between the active and inactive species, and at equilibrium most of the active material is present in the wash solution.

Volumetric operations

No mouth operations are permissible in a laboratory where unsealed radioisotope sources are handled. Mouth-operated wash bottles must not be used and pipettes must have suction devices to enable them to be operated other than by mouth. A selection of

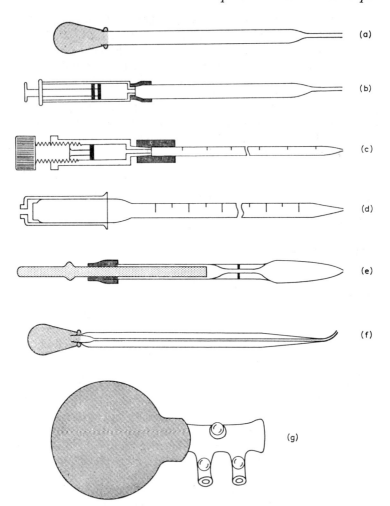

Fig 6.1 A selection of pipettes for radiochemical work

useful pipettes is illustrated in Fig. 6.1. Uncalibrated transfer pipettes, as shown at (*a*), are useful for liquid transfer operations; for example, removal of supernatant liquid from a precipitate in a centrifuge cone. An improved version of the transfer pipette may be made by connecting a disposable plastic syringe to the pipette with rubber tubing, as at (*b*). These syringes are calibrated from 0·2 to 2·0 cm³, so the pipette may be used for dispensing approximate volumes of solutions. A standard 1·0 cm³ calibrated pipette may be operated with a screw-type piston attachment, shown at (*c*). This

Practical Methods of Radiation Protection

gives a very precise control over the liquid level in the pipette and is particularly useful when wearing gloves as it does not depend upon making an air seal between the fingers and the pipette.

All-glass piston pipettes are available in capacities from 1·0 to 50 cm³ (Fig. 6.1(d)). In adjusting the meniscus the plunger must be held fully down and the hole closed with the finger. Some practice is required in using them whilst wearing gloves. A glass piston-operated 1·0 cm³ pipette such as that shown at (e) is useful where a number of dispensings of the same quantity of solution have to be made. For precise and rapid dispensing of small volumes, 0·01–0·25 cm³, the E-MIL auto zero pipette shown at (f) may be used. Suction bulbs (g) with a system of control valves may be used in conjunction with normal patterns of pipettes.

Micrometer-screw-operated syringes of the Agla type (Burroughs–Wellcome) are available for accurately dispensing very small volumes of solution, from 0·005 cm³ upwards. The tip of the needle of these syringes should be placed just under the surface of the liquid whilst the required volume is being delivered, and care must be taken to ensure that surplus solution on the outside of the needle does not contribute to the volume delivered. Such surplus solution may be removed by wiping the needle with a paper tissue.

Source preparation

In radioisotope work the relative activities of several materials have frequently to be measured. For this purpose it is necessary to prepare sources, most often from solutions of the materials, of precisely the same geometrical form for counting. These sources may be prepared on small watch-glasses, or on small metal counting trays, or planchettes. Metallic counting trays are available made from stainless steel, nickel-plated mild steel and aluminium. The choice between a glass or metal counting tray will depend upon the chemical nature of the solution to be evaporated.

The types of source commonly used are illustrated in Fig. 6.2. The source material, after evaporation in a flat-bottomed planchette (a), will be distributed evenly over the surface of the planchette. If only small volumes of liquid are to be evaporated, dimple planchettes (b), in which the liquid is confined to the dimple, will produce a source with a geometry which more nearly approximates to a point source. For convenience in handling these planchettes, they should be placed in watch-glasses before the solution is pipetted into them. Any carrier necessary should be added to the planchette before the active solution. The watch-glass and planchette may then be placed under an infra-red lamp for evaporation. Surface evaporation is carried out to avoid losses of active material such as would

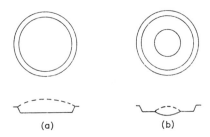

Fig 6.2 Source preparation

(a) Flat-bottomed planchette
(b) Dimple planchette

result from "bumping" in bottom-heat evaporation. For efficient evaporation the hood of the fume cupboard should be adjusted to maximize the flow of air across the surface of the planchette.

It is sometimes necessary to prepare sources from a solid material, usually a precipitate. The precipitate may be transferred as an aqueous or acetone slurry, using a transfer pipette, to a flat-bottomed planchette. The material should be distributed as evenly as possible so that consistent geometry may be obtained. If all the precipitate obtained during a chemical reaction is to be counted, the material may be collected on a glass-fibre filter disc using a demountable filter of the type shown in Fig. 6.3. By continual washing, the precipitate may be transferred quantitatively. The filter disc and precipitate are then transferred to a flat-bottomed planchette and dried before counting.

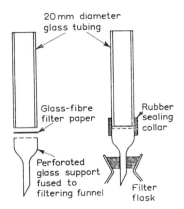

Fig 6.3 Demountable filter unit for the preparation of sources from precipitates

Practical Methods of Radiation Protection

Prepared sources should, whenever possible, be sealed to prevent loss of active material. The most convenient method of sealing a source after drying is by the addition and evaporation of a small amount of a 1% solution of cellulose acetate in acetone. This will seal the source with a thin layer of cellulose acetate which will have a negligible effect on the source counting rate for sources emitting energetic radiations. Sources of weak β-emitters, such as carbon-14 or sulphur-35, however, should not be sealed in this way as the sealing material would seriously absorb the particles emitted.

Decontamination of apparatus

All apparatus which has been in contact with active material must be assumed contaminated after use until properly monitored and proved to be free of active material. An end-window Geiger–Müller counter may be used for monitoring energetic β/γ-emitting isotopes, but for the weak β-emitters carbon-14 and sulphur-35, care will be required in interpreting any monitored activities. With these isotopes where there is extensive air and end-window absorption, activity levels will appear much lower than they really are. Alpha-emitters may be detected using an end-window Geiger–Müller counter, but in general both these and the very weak β-emitter tritium (hydrogen-3), require scintillation detectors for efficient monitoring. With the exception of the low-radiochemical-toxicity isotopes of thorium and uranium, neither tritium nor α-emitters in the open (unsealed) conditions have been used in the experiments in this book.

The following procedures may be used for the decontamination of apparatus, starting with the simplest method and progressing to the more severe methods until decontamination has been effected.

1. Wash with water
2. Wash with hot water and detergents
3. Wash in dilute acids
4. Wash in solution of complexing agents, e.g. ammonium citrate, ammonium bifluoride, ethylene diamine tetra-acetic acid (E.D.T.A.)
5. Wash in chromic acid

Methods 3 and 5 must obviously be omitted in the case of metal objects. For contamination on the hands or skin adopt one or other of the following procedures:

1. Brush with soap or detergent and warm water using a soft brush
2. Use an E.D.T.A. soap or wash with soap and E.D.T.A. solution
3. Immerse the hands in a saturated potassium permanganate

solution, rinse with water and immerse in 5% sodium bisulphite to remove the stain

Never continue hand or skin decontamination procedures to an extent where the skin becomes damaged.

Waste disposal

The control of radioactive waste disposal in the United Kingdom is the concern of the Department of the Environment. No definite levels for waste disposal can be given for any institution, for this will depend very much on local conditions. When disposal of liquid wastes into the drainage system is considered, it is necessary to ascertain what sources of drinking water may be polluted, and how many institutions there are in the vicinity that may dispose of radioactive material through the same system.

The implementation of the provisions of the Radioactive Substances Act (1960) concerning waste disposal is controlled by the Radiochemical Inspectorate of the Department of the Environment. It is possible, however, to give more precise guidance to institutions which work under the terms of the Radioactive Substances Act, Schools Exemption Order, i.e. to educational establishments that have elected to carry out work within the specified activity limits of this Order. This will be discussed in Chapter 7.

Dealing with accidents

In accidents which involve personal injury, the treatment of the injury must always take precedence over decontamination. For serious injuries medical advice must of course be sought, but first aid treatment should be carried out as if radioactivity were not involved. In the case of minor cuts or scratches, the wound should be thoroughly irrigated with water to reduce the possibility of entry of radioactive substances into the bloodstream.

The commonest form of laboratory accident which will be met, however, is the spillage of radioactive materials. With proper laboratory practice such spills should in all cases be contained within a limited area; for example, within the confines of the working tray. For a wet spill, copious quantities of paper tissues should be used to soak it up, and these may then be discarded as active solid waste. The area should then be monitored, and if the contamination is considered to be excessive, work should be discontinued there until decontamination has been effected by more rigorous methods. Solid spills must first be moistened to prevent spread of contamination by air currents. After moistening the solid it can be mopped up with tissues. In clearing up any spill it is important to limit the spread of the radioactive material as much as possible.

GENERAL LABORATORY REGULATIONS

It is important that everyone working with open sources of radioisotopes should understand the main provisions which were discussed above. As a guide for those who are initiating such work the following laboratory rules are most important.

1. No eating, drinking or mouth operations of any kind are allowed in radioisotope laboratories.
2. All cases of personal injury, however trivial, must be reported to a member of staff.
3. Laboratory coats must be worn at all times.
4. Protective gloves must be worn when handling radioisotopes as open sources.
5. All radioactive work must be conducted over a contained area lined with impervious paper.
6. When not required, stock radioisotope solutions should be kept in a suitably locked cupboard. Bottles containing these solutions should be placed in leak-proof metal cans and the intervening space fitted with absorbent material.
7. Any active spill must be dealt with and reported immediately.
8. All glassware must be washed and monitored after use. On no account must contaminated glassware be stored with non-contaminated apparatus.
9. All labelled materials or contaminated articles that are left in the laboratory must be adequately marked.
10. Liquid waste must not be poured into sinks by unauthorized persons. This must be carried out by a member of staff who must keep the necessary records. (see Chapter 7.)
11. The hands must be washed on leaving the laboratory.

Chapter 7

Regulations for the use of radioisotopes in educational establishments

The Department of Education and Science (D.E.S.) have issued regulations concerning the use of radioisotopes and apparatus which produces ionizing radiations in schools and colleges within the United Kingdom [1]. These regulations apply to all schools and to those higher educational establishments which have elected to work at activity levels exempting them from the provisions of the Radioactive Substances Act (1960) [2]. The regulations require that schools or colleges intending to carry out work under this Schools Exemption Order (1963) [3] should notify and seek the approval of the D.E.S. before doing so.

The regulations state that the maximum permitted dose that a pupil or student may receive as a result of experimental work with radioactive materials is 50 mrem/year. This figure applies only to external radiation exposure. This has been adopted as a safe dose level since it forms only a small fraction of the ICRP recommended genetic dose to individuals in the population of 2 rems up to the age of 30 years (see Chapter 6). The aim of the D.E.S. was to set a dose limit for young persons that provides a considerable degree of protection against possible effects of radiation.

It is useful to compare this level of 50 mrem/year with some activities of β/γ-emitting isotopes which, at a distance of one foot, contribute a dose of 1 mrem/hour. These are

100 μCi	Radium-226
70 μCi	Cobalt-60
300 μCi	Caesium-137

A student would have to spend 50 hours at one foot from these sources during a year to accumulate the quoted maximum dose. Not only is this extremely unlikely, but the activities quoted above are far higher than are necessary for most experimental purposes. The average activity of the closed sources used in the experiments described later ranges from 5 to 10 μCi. In fact the maximum activity of any such source must not generally exceed 10 μCi, and should higher activities be required for particular purposes, special permission must be sought from the D.E.S.

Regulations for Use of Radioisotopes

Throughout the foregoing discussions on radiation protection, maximum permissible levels of dose in rems have been quoted extensively. Such levels tend to be somewhat confusing unless some comparisons with natural radiation levels can be made. It is interesting, for this reason, to quote a range of background levels that populations are exposed to in certain regions of the world. In the United Kingdom, the average dose-rate due to cosmic radiation and the natural radioisotopes present in the rocks and soil is of the order of 100 mrem/year. In predominantly limestone areas such as the Mendip Hills in Somerset it is 40 mrem/year, and in granite areas such as parts of Cornwall it can be of the order of 120 mrem/year. In the Brazilian state of Rio de Janiero, about 50 000 people live on thorium-bearing soils that contribute a dose of 1000 mrem/year. Elsewhere in Brazil, doses of the order of 12 000 mrem/year have been reported. In the state of Kerala in India about 100 000 people are exposed to doses of 2600 mrem/year. Not only is the external dose to these people considerably higher than average, but the food they eat contains traces of the elements that contribute to these high external doses. It can therefore be concluded that a dose of 50 mrem/year, a dose which under normal conditions is difficult to accumulate in educational courses, is very insignificant compared with many background levels of radiation. It may be regarded as a very safe level of exposure.

Although the external doses from microcurie-level sources are extremely small, it is nevertheless useful to know what they are, so that any hazard may be assessed. The dose-rates from such sources may be either calculated from the formulae given in Chapter 6, or alternatively measured by a dose-rate meter. Table 12 gives the results of some experimentally determined dose-rates for sources approved by the D.E.S. It may be noted from the table that the γ-ray dose-rate from sources such as cobalt-60 are little attenuated by the small amount of lead shielding incorporated in the storage box. The main purpose of the box is to prevent loss of the source. The energetic β-particles from a strontium-90 source, which in close proximity produce a high β-dose rate, are completely absorbed by the storage container. The γ-ray dose-rates from these sources are generally smaller than those from a bottle of a uranium or thorium salt, and at distances greater than 30 cm are at most only about four times background. It will obviously be relatively easy to ensure that no pupil receives a dose exceeding 50 mrem/year as a result of experiments using such sources.

The units quoted in Table 12 are expressed in multiples of röntgens per hour, since the instruments used for their measurement measure only the exposure dose. Because we are dealing with β- or γ-radiation with a quality factor of 1, 1 R = 1 rad = 1 rem. Thus

Table 12 Dose rates due to some Department of Education and Science approved sources

Source	Distance from ratemeter, cm	γ-ray dose-rate (including background), $\mu R/h$
1. Background		8
2. Cobalt-60, 5 μCi		
(a) Out of box	10	180
	30	35
	50	20
	100	12
(b) In lead pot in storage box for single source	10	150
	30	32
	50	18
	100	11
3. Americium-241, 5 μCi		
(a) Out of box	As close as possible to ratemeter	30 (γ-activity)
(b) In box	ditto	(background only)
4. Radium-226, 5 μCi		
(a) Out of box	10	150
	30	32
	50	18
	100	12
(b) In box as above	10	70
	30	18
	50	12
5. Collection of 3 boxes of sources: (a) Single source Co-60 as above (b) Single source Am-241 as above (c) Box of 3 sources Ra-226 as above Sr-90, 1 μCi Pu-239, 0·1 μCi	50	24 The dose rate just outside a storage cupboard from such a collection of sources placed at the back of the cupboard would therefore be 2–3 times background rate
6. Thorium nitrate, 500 g in glass bottle	10	350
	30	70
	50	32
	100	16
7. Uranyl nitrate, U-238 enriched, 250 g in glass bottle	10	80
	30	24
	50	17
	100	14
		β-particle dose rate, mR/h
8. Strontium-90, 1 μCi Out of box	As close as possible to a β-monitor	80
	15	0·5

Regulations for Use of Radioisotopes

for protection purposes the dose from such sources can be assumed to be in units of rems/hour.

In the use of open sources of isotopes, the D.E.S. regulations ensure that no pupil or student will be exposed to high internal contamination risks by severely limiting the activity levels of isotopes that may be used for each experiment. This maximum level of activity for an isotope is one-tenth of the ICRP limit for ingestion of the isotope by an occupational worker in one year. These maximum permissible body burdens were discussed in Chapter 5 and are listed in Table 9. If we consider iodine-131 as an example, the maximum

Table 13 Maximum quantities of isotopes that may be used per experiment in schools

Isotope	Maximum quantity per experiment, μCi
Sodium-24	160
Potassium-42	240
Phosphorus-32	16
Sulphur-35	48
Chlorine-36	64
Carbon-14	640

permissible concentration in water (MPC$_w$) for a 168 hour week is quoted as 2×10^{-5} μCi/cm^3. A "standard man" drinks 2200 cm^3 of water daily, so he would take into his body during one year

$$2 \times 10^{-5} \times 2200 \times 365 = 16 \,\mu\text{Ci}$$

The maximum permissible quantity of iodine-131 for one experiment is, then, one-tenth of this, or 1·6 μCi. The total activity for one isotope that may be stored on school or college premises is five times the maximum permissible quantity per experiment, which for iodine-131 is 8 μCi. Table 13 quotes these quantities for other isotopes which are used as open sources in experiments described later. Although in general five times the maximum permissible quantity per experiment may be kept as stock solutions, an overriding limit is set by the regulations in that the total activity of all open sources held in stock should not exceed 2 mCi of activity. The total activity allowed for both open and closed sources that may be held in stock is 4 mCi. If more than one radioisotope is to be used in a single experiment, then the activities of each must be reduced proportionately. For example, in Experiment 29.2, both phosphorus-32 and sulphus-35 are used together, and therefore the maximum activities to be used in this case would be 8 μCi of phosphorus-32 and 24 μCi of sulphur-35. The use of the more hazardous radioisotopes, namely those of the high-toxicity classification (which include the α-emitters and strontium-90) as open sources is prohibited in schools. They may, however, be used as sources in the closed form.

The D.E.S. regulations require that no pupils under the age of 16 years be permitted to handle radioactive materials, other than natural uranium, thorium or potassium compounds. They should not be allowed into laboratories where older pupils are conducting experiments with radioisotopes, although the demonstration of such experiments by the teacher is allowed.

WASTE DISPOSAL

According to the Radioactive Substances Act, School Exemption Order 1963, schools may dispose weekly of 500 μCi of liquid waste into the main drainage system, and 10 μCi of solid waste through the local authority's refuse collection service. If the waste is gaseous (Experiment 27) a maximum activity of 1 μCi/day may be released into the atmosphere. Because of the higher levels of activity that are permissible as liquid waste, it is a good policy to convert as much of the solid or gasous waste as possible to the liquid or solution form and dispose of it into the drains together with a large volume of water to effect dilution. Waste material should not be stored, since this would constitute a hazard, but should be disposed of as soon as possible.

The Exemption Order, mentioned above, requires that records of radioactive waste disposals be kept. These must show the date of disposal, the activity and nature of the isotope and the form of the waste disposal. These records must be kept available for inspection by the Radiochemical Inspectorate, if requested.

It must be emphasized that the above is a summary of the main provisions of the D.E.S. regulations. Any teacher who intends to use open or closed sources of radioisptopes should obtain a copy of the regulations and familiarize himself with their detailed requirements.

REFERENCES

1   *The Use of Ionising Radiations in Schools, Establishments of Further Education and Teacher Training Colleges*, Administrative Memorandum 1/65 (Department of Education and Science 1965)
2   *Radioactive Substances Act*, 1960 (London, HMSO)
3   *Radioactive Substances (Schools etc) Exemption Order*, 1963 (London, HMSO)
4   *Radiation Protection in Schools for Pupils up to the Age of 18 years*, ICRP Publication 13 (Pergamon Press 1970)

Experiment 1
The determination of half-life

Radioactive decay is a random process involving the individual nuclei of a radioisotope. The probability that a particular radioactive nucleus will decay in any selected time interval is independent of the state of neighbouring nuclei, the chemical state of the atom and physical conditions such as temperature and pressure. In a collection of radioactive nuclei, those which disintegrate in unit time are randomly distributed and the decay process may be treated statistically. Each nucleus of a given radioactive species has the same finite probability, λ, of disintegrating in any unit time interval. If the number of such nuclei which are present at time t is N, then the number of nuclei which will disintegrate in the interval from t to $t + dt$ is given by

$$dN = -\lambda N \, dt$$

or

$$\frac{dN}{dt} = -\lambda N \tag{E.1.1}$$

where dN/dt is the disintegration rate, or activity, and λ is the decay constant characteristic of the particular radioisotope. Integration of this equation gives

$$\log_e N = -\lambda t + C$$

where C is a constant.

Substituting $N = N_0$, the initial number of nuclei present at $t = 0$, shows that the constant C has the value $\log_e N_0$. Therefore

$$\log_e \frac{N}{N_0} = -\lambda t$$

or

$$N = N_0 e^{-\lambda t} \tag{E.1.2}$$

The activity, A, of a radioactive preparation is defined as its decay rate, and since $dN/dt \propto N$, eqn. (E.1.2) may be rewritten as

$$\frac{dN}{dt_t} = \frac{dN}{dt_0} e^{-\lambda t} \tag{E.1.3}$$

or

$$A_t = A_0 e^{-\lambda t} \tag{E.1.4}$$

and

$$\log_e A_t = \log_e A_0 - \lambda t \tag{E.1.5}$$

This exponential decrease of activity is illustrated graphically in Fig. 1.9. A graph of $\log_e A$ against t will be linear since eqn. (E.1.5) is the equation to a straight line.

THE HALF-LIFE OF RADIOISOTOPES

The half-life of a radioisotope, $t_{1/2}$, is defined as the time required for one-half of the radioactive nuclei present at any time to decay. Substituting $N/N_0 = \frac{1}{2}$ when $t = t_{1/2}$, in eqn. (E.1.2), gives the following relationship between $t_{1/2}$ and λ:

$$t_{1/2} = \frac{\log_e 2}{\lambda} = \frac{0 \cdot 693}{\lambda} \tag{E.1.6}$$

If the half-life of a radioisotope lies between a small fraction of a second and several months, then it is possible to determine the half-life by making a series of activity measurements over a period of several half-lives.

Changing the base of the logarithms in eqn. (E.1.5),

$$\log_{10} A_t = \log_{10} A_0 - \frac{\lambda t}{2 \cdot 303} \tag{E.1.7}$$

so that the straight line obtained on plotting $\log_{10} A_t$ against time will have a slope of $\lambda/2 \cdot 303$. An accurate value for the half-life may therefore be calculated from the slope of such a graph.

If the half-life is longer than a few months it becomes impracticable to measure the change in activity as a function of time. Eqn. (E.1.1) may be written

$$\lambda = -\frac{dN/dt}{N} = -\frac{\Delta N/\Delta t}{N} \tag{E.1.8}$$

where ΔN is the number of disintegrations occurring in the finite time interval Δt, which is small compared to the length of the half-life. The number of atoms, ΔN, present during this interval may be assumed constant and can be calculated from the mass of the sample of radioisotope, which may have to be determined by chemical analysis.

It must also be noted that it is necessary to determine absolutely the number of disintegrations, ΔN, which occur in the time Δt.

Experiment 1

Detection methods which record only a fraction of the total disintegrations, for example Geiger–Müller counters, are therefore not suitable.

Experimental

It is suggested that the mathematical form of the radioactive decay law may with advantage first be studied using the water-flow analogy described in Experiment 2. An elaboration of the water analogy allows more complex radioactive decay phenomena such as the decay of mixtures of radioisotopes or radioactive equilibria to be studied. For the study of radioactive equilibria the analogy has the advantage not only that it is a visual demonstration but also that the principles involved do not become obscured by difficult experimental procedures which are necessary in studying equilibria quantitatively. After studying some decay processes using the water analogy, the following experimental work which involves the actual determination of half-lives may be undertaken.

Those readers who require only to understand the kinetics of simple radioactive decay should carry out Experiment 19 (Method 2) and determine the half-life of protoactinium-234. The chemical separation involved is very easy, and a graph similar to that shown in Fig. 1.9 can readily be obtained. The half-life may be read directly from this graph, but it is more accurate to determine it from the slope of a $\log_{10}$(activity)/time plot as described above.

(A) Half-life determinations involving the measurement of activity change with time

The following experiments which are described later involve the determination of half-life by the measurement of the variation of activity with time. The following points should be noted in making such measurements:

1. It is assumed in measuring the counting rate that the average rate observed is the true rate at the mid-point of the counting period. This is in fact only a good approximation if the length of the counting period does not exceed one half-life of the radioisotope.
2. Provided that the above rule is not violated, the counting interval should be long enough to record at least 10 000 counts for each activity determination (see Experiment 5).
3. The counting rates must be corrected for counter paralysis time and background. The half-live may then be obtained from a plot of $\log_{10}$(counting rate) against the average time at which each count was taken.

Experiment 18. The separation of radon-220 from the thorium series and the determination of its half-life using an ionization chamber

Experiment 19. The radiochemical separation of protoactinium-234 from the uranium-238 series

Experiment 20. The radiochemical separation of lead-212 and bismuth-212 from the thorium series

Experiment 21. The radiochemical separation of praseodymium-144 from cerium-144

Experiment 22. The preparation of iodine-128 using a laboratory neutron source

(B) Half-life determinations involving the measurement of absolute activity

The half-life of uranium may be determined by measuring the number of disintegrations occurring in a finite time interval using nuclear emulsion plates; this is described in Experiment 23.

Experiment 2

Radioactive decay phenomena simulated by the flow of water through tubes

The rate processes involved in radioactive decay may be simulated by the viscous flow of water through a capillary tube. In the arrangement shown in Fig. E.2.1 for streamline flow of a liquid through the capillary, the rate of change of the volume in the large tube at any time is given by Poiseuille's equation:

$$\frac{dV}{dt} = - \frac{\pi h \rho g a^4}{8 \eta l} \quad \text{(E.2.1)}$$

where V = Volume of liquid in large tube at that time
h = Height of liquid in large tube at that time
l, a = Length and radius of capillary
ρ, η = Density and viscosity of liquid

The only variable on the right-hand side of this equation is h, and since V is proportional to h, we may write

$$\frac{dh}{dt_t} \propto -h_t \quad \text{(E.2.2)}$$

or

$$\frac{dh}{dt_t} = -\lambda h_t \quad \text{(E.2.3)}$$

where λ is a constant. This equation is obviously analogous to the radioactive decay law (see Experiment 1), h_t being the equivalent of N_t, the number of radioactive atoms present at time t. Measurements of dV/dt or dh/dt, the rate of change of the volume or height of water in the large tube, are formally the same as measurements of dN/dt, the decay rate of a radioisotope. The "half-life" of a tube may be observed visually and can be varied by altering the capillary radius or length. Some approximate dimensions and "half-lives" are given in Fig. E.2.1.

It is important that the flow should obey Poiseuille's law, and for this to be so there must be no net gain in kinetic energy due to gravitational acceleration as the water flows through the capillary. To ensure that this condition is fulfilled it is essential for the capillary

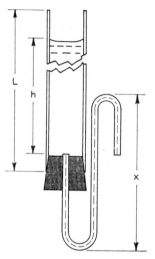

Fig E.2.1 Tube with capillary outlet for the simulation of decay phenomena by water flow

Apparatus dimensions:
Large tube: length, L, 65 cm, diameter, 3 cm
The half-life for a tube of these dimensions will vary with capillary tube dimensions as follows:

Half-life, min	Capillary dimensions	
	diameter, mm	length, x, cm
2·5	1·5	15
4	1·5	24
5	1·25	16
7·5	1·25	24
12	1·0	16

tube to be bent to the shape shown in Fig. E.2.1, with the inlet and outlet at the same horizontal level. The following varieties of tube may be used in various combinations to simulate decay phenomena as described below.

(A) A tube with a capillary outlet and a constant-level device. This simulates an isotope with a very long half-life.
(B) A tube with a long capillary outlet to simulate a long-lived isotope but of shorter half-life than A.
(C) A tube with a short capillary outlet to simulate an isotope of short half-life.
(D) A closed tube to simulate a stable isotope.

The decay phenomena which can be studied with these tubes are:

(*a*) The decay of a radioisotope and the determination of its half-life.

Experiment 2 109

(b) The decay of a mixture of radioisotopes and the determination of the half-lives of the components.

(c) Radioactive equilibria involving the growth and decay of radioactive daughter isotopes.

(a) THE SIMULATION OF SIMPLE DECAY

We have seen in Experiment 1 that for radioactive decay

$$N_t = N_0 e^{-\lambda t} \tag{E.2.4}$$

where N_t is the number of atoms of the radioisotope present at time t, and N_0 is the number initially present. The volume or height, h_t, of water in the large tube is the analogue of N_t.

Experimental

1. Allow water to flow from tube B (or C) into tube D. Record the height of the water level (h) in tubes B and D at intervals of about 2 min.
2. Plot h_B and h_D against time to obtain exponential curves for the decay of the parent, B, and growth of the stable daughter, D.
3. Plot $\log_{10} h_B$ against time and determine the half-life of B from the straight line obtained.

(b) THE SIMULATION OF THE DECAY OF A MIXTURE OF RADIOISOTOPES

A mixture of radioisotopes of species 1 and 2 will decay according to the equations

$$N_1 = N_{01} e^{-\lambda_1 t} \tag{E.2.5}$$

and

$$N_2 = N_{02} e^{-\lambda_2 t} \tag{E.2.6}$$

The total number, N_t, of radioactive atoms remaining after time t will be given by

$$N_t = N_1 + N_2 = N_{01} e^{-\lambda_1 t} + N_{02} e^{-\lambda_2 t} \tag{E.2.7}$$

After sufficient time has elapsed for one of the isotopes, say species 1, to decay completely,

$$N_t = N_{02} e^{-\lambda_2 t} \tag{E.2.8}$$

so that the latter part of a plot of $\log_{10} N_t$ against t will be a straight line from which the half-life of species 2 can be obtained. The extrapolated straight line which represents the decay of species 2 may then be subtracted from the composite decay curve to yield a

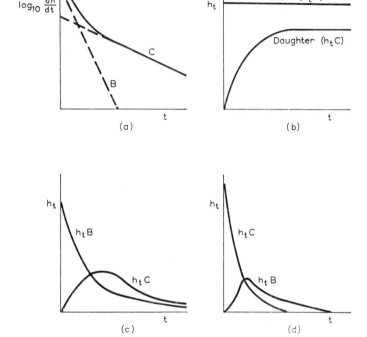

Fig E.2.2 Radioactive decay phenomena

(a) Decay of a mixture of radioisotopes
(b) Secular equilibrium, $\lambda_1 \ll \lambda_2$
(c) Transient equilibrium, $\lambda_1 < \lambda_2$
(d) $\lambda_1 > \lambda_2$

straight line from which the half-life of species 1 can be determined. This is illustrated in Fig. E.2.2(a).

Experimental

1. Allow water to flow simultaneously from tubes B and C into a closed tube D, which should be large enough to hold the total contents of B and C. Record the liquid level in D every 2 min.

2. Plot $\log_{10}(h_n - h_{n-1})$ against time, where h_n is the nth observation of the level in D. This value is analogous to the decay rate—the quantity that would be measured in an experiment using radioisotopes.

3. From the straight-line portion of this graph determine the half-life of B. Extrapolate the straight line to zero time and subtract it from the total curve and hence determine the half-life of C.

(c) THE SIMULATION OF RADIOACTIVE EQUILIBRIA

Radioactive equilibria occur in those cases where the product of the decay of a radioisotope is itself radioactive as in the decay sequence:

Species 1 → Species 2 → Species 3
(radioactive parent) (radioactive daughter) (stable)

The rate of decay of the parent species is given by the equation

$$\frac{dN_1}{dt} = -\lambda_1 N_1 \tag{E.2.9}$$

The decay rate of the daughter, if it is isolated from its parent, is similarly given by

$$\frac{dN_2}{dt} = -\lambda_2 N_2 \tag{E.2.10}$$

However, in the presence of the parent species, atoms of species 2 are continually being formed by decay, and the net decay rate of species 2 is given by

$$\frac{dN_2}{dt} = \lambda_1 N_1 - \lambda_2 N_2 \tag{E.2.11}$$

$$= \lambda_1 N_{01} e^{-\lambda_1 t} - \lambda_2 N_2 \tag{E.2.12}$$

The solution of this equation for N_2, the number of atoms of species 2 which are present at time t, is

$$N_2 = \frac{\lambda_1}{\lambda_2 - \lambda_1} N_{01}(e^{-\lambda_1 t} - e^{-\lambda_2 t}) \tag{E.2.13}$$

Three kinds of radioactive equilibria may now be distinguished according to the relative magnitudes of the radioactive decay constants of the parent and daughter species, λ_1 and λ_2.

(i) Secular equilibrium

If the half-life of the parent is very much longer than that of the daughter species, $\lambda_1 \ll \lambda_2$ and eqn. (E.2.13) becomes

$$N_2 = \frac{\lambda_1}{\lambda_2} N_{01}(e^{-\lambda_1 t} - e^{-\lambda_2 t}) \tag{E.2.14}$$

Since $\lambda_2 t = (0 \cdot 693/t_{1/2})_2 t$, after several half-lives of the daughter have elapsed since the parent started to decay, $e^{-\lambda_2 t}$ approaches zero and eqn. (E.2.14) becomes

$$N_2 = \frac{\lambda_1}{\lambda_2} N_{01} e^{-\lambda_1 t} = \frac{\lambda_1}{\lambda_2} N_1$$

or

$$\lambda_2 N_2 = \lambda_1 N_1 \tag{E.2.15}$$

Once this condition is satisfied secular equilibrium has been established and the decay rates of parent and daughter are the same.

Secular equilibrium is illustrated in Fig. E.2.2(b). It may be noted that, since the decay rate of the parent is very slow compared with that of the daughter, the number of atoms of the parent present remains approximately constant. The number of atoms of the daughter increases over a period of several times its half-life to a constant value. The time required for this equilibrium to be established (so that N_2 reaches 97% of its final constant value) is equal to 5 half-lives of the daughter, species 2.

Secular equilibrium is probably the most important of the radioactive equilibria. It occurs frequently amongst the natural radioelements as well as with some pairs of artificial radioelements. Several experiments described later involve the separation of elements in secular equilibrium.

Experimental

1. Allow water to flow in the following sequence:

   tube A → tube C → tube D

Tube A should initially be full and tubes C and D should be empty. Record the liquid levels (h) in tubes C and D every 2 min.

2. Plot h_C and h_D against time. The level in tube C will become constant as equilibrium is reached. The rate of change of the level in tube D will become constant after equilibrium has been reached. This constant rate of change corresponds to the constant decay rate of C, which then equals the decay rate of A.

3. Determine the time required to reach equilibrium in terms of the half-life of C.

(ii) Transient equilibrium

If the parent species is longer lived than the daughter so that $\lambda_1 < \lambda_2$, then eqn. (E.2.13) must be applied without approximation. Fig. E.2.2(c) illustrates the decay of the parent and growth of the daughter for this case. The maximum in the number of daughter atoms present must occur when the decay rate of the daughter is just equal to its rate of formation by decay of the parent species.

Experiment 2

The time, t_m, required to reach this maximum may be obtained by making the derivative of eqn. (E.2.13) zero, whence

$$t_m = \frac{1}{\lambda_2 - \lambda_1} \log_e \left(\frac{\lambda_2}{\lambda_1}\right) \quad \text{(E.2.16)}$$

After several half-lives of the daughter have elapsed, the decay curves for the daughter and parent become parallel and the daughter decays with the same half-life as the parent. At this part of the decay curves there is a "transient equilibrium" in which the ratio of the numbers of parent and daughter atoms is constant. Once this equilibrium has been reached $e^{-\lambda_2 t}$ is negligible compared to $e^{-\lambda_1 t}$ and eqn. (E.2.13) becomes

$$N_2 = \frac{\lambda_1}{\lambda_2 - \lambda_1} N_{01} e^{-\lambda_1 t} = \frac{\lambda_1}{\lambda_2 - \lambda_1} N_1$$

or

$$N_2/N_1 = \frac{\lambda_1}{\lambda_2 - \lambda_1} \quad \text{(E.2.17)}$$

Experimental

1. Allow water to flow in the following sequence:

   tube B → tube C → waste

Initially, tube B should be full and tube C empty. Record the water levels in tubes B and C every 2 min.

2. Plot (*a*) h_B and h_C against time, and (*b*) $\log_{10} h_B$ and $\log_{10} h_C$ against time.

3. Determine the half-life of the parent, and hence λ_1, from the plot of $\log_{10} h_B$ against time.

4. From the time required to reach the maximum of the plot of h_C against time, calculate the half-life of the daughter using eqn. (E.2.16).

5. From the transient equilibrium region of the plots of h_B and h_C against time, determine the equilibrium ratio N_2/N_1 and calculate the half-life of the daughter using eqn. (E.2.17).

(iii) Half-life of parent is shorter than half-life of daughter

If a radionuclide decays to a radioisotope of shorter half-life, the number of atoms of the daughter present at time t is again given by eqn. (E.2.13). Fig. E.2.2(*d*) illustrates this phenomenon. As in the case of transient equilibrium, the time, t_m, required to reach the maximum activity of the daughter can be obtained by making the derivative of eqn. (E.2.13) equal to zero. This gives the following equation for t_m:

$$t_m = \frac{1}{\lambda_1 - \lambda_2} \log_e \left(\frac{\lambda_1}{\lambda_2}\right) \tag{E.2.18}$$

In this case, after the maximum in the daughter activity has been reached, the parent decays more rapidly than the daughter until, finally, only the daughter remains.

Experimental

1. Allow water to flow in the sequence

    tube C → tube B → waste

Initially tube C should be full and tube B empty. Record the water levels in tubes C and B every 2 min.

2. Plot (*a*) h_C and h_B against time, and (*b*) $\log_{10} h_C$ and $\log_{10} h_B$ against time.

3. Determine the half-life of the parent, and hence λ_1, from the plot of $\log_{10} h_C$ against time.

4. From the time required to reach the maximum of the plot of h_B against time, calculate the half-life of the daughter using eqn. (E.2.18).

5. The half-life of the daughter may also be obtained from the tail of the plot of $\log_{10} h_B$ at times corresponding to complete decay of the parent.

Experiment 3

Electroscopes and ion chambers for radiation detection

The electroscope is an instrument for measuring quantity of charge. The simple electroscope, such as those illustrated in Fig. E.3.1 consists of a thin leaf of gold or aluminium, attached to a rigid metal support. The whole leaf assembly is insulated from its surroundings and earth and is enclosed in a draught-proof container. If the leaf assembly is charged, the leaf will be repelled from its support since both the leaf and the support must carry charge of the same sign. The leaf will gradually return to the uncharged position if the assembly subsequently loses charge. This happens when the air path to earth is made conducting by the ionizing effect of a radioactive source placed nearby. Increasing the activity of the radioactive source increases the rate of fall of the leaf.

It is possible to use electroscopes to investigate α-particle and β-particle absorption. For α-particle absorption, air may be used as the absorbing medium and the discharge rate of the electroscope determined for various air distances between the source and the electroscope window. For this experiment the α-particle source need only have an activity of a few microcuries. For quantitative work, however, it is more satisfactory to use a combination of a pulse electroscope and an ion chamber such as that described below. The sensitivity of ion chambers for β-particles is very much less than their sensitivity for α-particles due to the lower specific ionization of β-particles. Their use for β-particle absorption experiments therefore requires source activities of at least 50 to 100 μCi, and some caution is necessary in handling such sources. It is better to study β-particle absorption using a Geiger–Müller counter as described in Experiment 7. A source activity of about 1 μCi is then sufficient.

The pulse electroscope can be used as a current-measuring device. It is generally used together with an ionization chamber for radioactivity measurements, and its main features are shown in Fig. E.3.2 and Plate 7. More elaborate pulse electroscopes are available in which an image of the leaf assembly can be projected.

The ionization chamber, 5, is held at a constant potential which is high enough to ensure that any ions formed in the chamber will be collected at the electrodes. The central electrode, 2, carries a fine

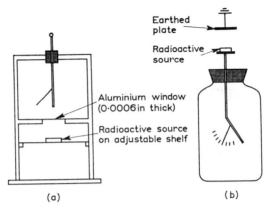

Fig E.3.1 Examples of the simple electroscope
(a) End-window electroscope
(b) "Bottle" electroscope

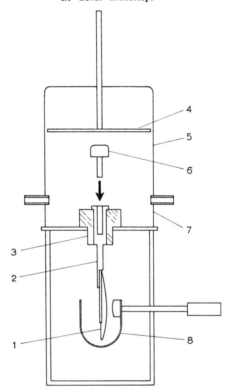

Fig E.3.2 Ion chamber of adjustable size with a pulse electroscope

1. Leaf
2. Central electrode
3. Insulator
4. Adjustable plate
5. Ionization chamber
6. Radioactive source
7. Positive high voltage via 10 MΩ safety resistor
8. Earth electrode

aluminium foil which is hinged at its upper end and tied back to the electrode at its lower end by a flexible attachment. This attachment may be a quartz fibre, or the leaf itself may simply be bent round and fastened at the back of the electrode.

If a radioactive source is placed in the ionization chamber, ionization of the air occurs and the current flows between the outer case of the chamber and the central electrode. As this electrode acquires charge the aluminium foil, having charge of the same polarity, is repelled. There will come a point, as the charge increases, at which the aluminium foil is repelled sufficiently to touch the earthed electrode, 8. The central electrode then loses its charge, returns to its original position, and the cycle is then repeated. One such cycle, or discharge, corresponds to a fixed amount of charge transferred from the ionization chamber case to the central electrode. The earthed electrode is movable, and the closer it approaches the central electrode, the smaller will be the amount of charge transferred in the cycle. The number of such discharges occurring per second is a measure of the rate of charge transference to the central electrode, i.e. of the ionization-chamber current, which in turn is a measure of the activity of the radioactive source. If the distance between the central electrode and the earthed electrode is small, then low activities may be measured.

THE CURRENT IN AN IONIZATION CHAMBER

The current due to the collection of the ions produced when a radioactive source causes ionization in an ion chamber may be calculated as follows. Consider a 5 μCi source which emits 5 MeV α-particles, placed inside the ion chamber. Suppose that the α-particles are totally absorbed in the air within the chamber. These conditions would apply to the experiments with pulse electroscopes and ion chambers described later.

The energy required to form an ion pair in air at standard temperature and pressure is about 32·5 eV. The total absorption in air of the kinetic energy of a 5 MeV α-particle will therefore result in the production of approximately 154 000 ion pairs. The disintegration rate of a 5 μCi source is 18·5 × 10^4 disintegrations/second, and the source construction will generally permit about half of these to enter the ion chamber. The number of ion pairs formed in the chamber in a second is therefore

$$154\,000 \times \frac{18 \cdot 5}{2} \times 10^4 = 1.43 \times 10^{10} \text{ ion pairs/second}$$

Since the charge on the electron is 1·6 × 10^{-19} C, the current resulting from the collection of these ion pairs in the ion chamber will be

$$1.43 \times 10^{10} \times 1.6 \times 10^{-19} = 2.28 \times 10^{-9} \text{ A}$$

This is the maximum current which will arise due to the source, and it may be reduced if some of the α-particles are prevented from entering the ion chamber due to the source construction or if they dissipate their energy by collision with the walls of the ion chamber itself. A current-measuring device with a sensitivity of about 10^{-12} A can therefore be used with an ion chamber to measure α-particle sources with activities of about 1 nCi, i.e. about 2000 disintegrations/minute, or more. Most pulse electroscopes can detect currents of this order, but measurements are much easier with sources of about 1 μCi.

In the case of β-particles, each particle will generally lose only a fraction of its total energy within the ion chamber because of the much greater range of β-particles in air. It must also be remembered that the average energy of the β-particles is only about one-third of the maximum (see Experiment 7). A 5 μCi source emitting β-particles of maximum energy equal to 1 MeV, for example, would produce an ion chamber current of the order of 10^{-12} to 10^{-14} A in the ion chambers used with pulse electroscopes, which is below their limit of sensitivity.

Experimental: Current/voltage characteristics of an ionization chamber using a pulse electroscope

If a voltage is applied across an ionization chamber in which the air is ionized by, for example, radiation from a radioactive source, the ionization current will increase with increasing voltage until it reaches a constant value as in Fig. E.3.3. This constant value is the saturation ionization current and is obtained when all the ion pairs formed are collected at the electrodes. At lower voltages some of the

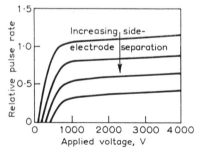

Fig E.3.3 Variation of the pulse rate of a pulse electroscope with ionization-chamber potential and side-electrode separation

Experiment 3 119

ion pairs recombine before they can be collected, resulting in a lower ionization current.

A pulse electroscope can be used to measure the ionization current in terms of the discharge rate of the leaf. This discharge rate is proportional to the ionization current for a given separation between the earthed electrode and the central electrode (Fig. E.3.2).

1. Place a 5 μCi α-particle source on the central electrode of the ion chamber.
2. Set the side electrode separation to the minimum distance at which satisfactory operation of the pulse electroscope can be obtained when a potential of about 2000 V is applied to the ion chamber, as indicated in Fig. E.3.2. The 10 MΩ safety resistor should be located at the outlet of the power supply so that all parts of the electroscope can be handled safely. This resistor is also essential to prevent damage to the leaf assembly in the event of a short-circuit.
3. Determine the discharge frequency of the leaf for a series of voltages from zero to about 3000 V.
4. Repeat the measurements for other side-electrode separations.
5. Plot the discharge frequency against voltage for each side-electrode separation used. The voltage at which the saturation ion current can be measured may now be obtained from the graph for each side-electrode setting.

Experiment 4
The Geiger–Müller counter

EXPERIMENT 4.1 PLATEAU CURVE AND EFFICIENCY OF A GEIGER–MÜLLER COUNTER

The gas ionization phenomena which take place in a Geiger–Müller counter have been discussed in Chapter 3. As the anode voltage is raised, the gas amplification factor and hence the pulse size from the counter increase and the latter eventually becomes large enough to operate the scaler, which then begins to register counts. On further increasing the anode voltage the situation is rapidly reached where all the particles or photons entering the counter produce pulses large enough to be recorded by the scaler. The relationship between counter anode voltage and the counting rate of a source, known as the *plateau curve* of the counter, is shown in Fig. E.4.1. At the starting voltage, V_s, radiations which enter the counter axially can produce pulses large enough to actuate the scaler. At the threshold voltage, V_t, radiations entering the counter from all directions produce pulses, all of the same size, large enough to be counted. Between these two voltages the pulse size produced by a particle or photon depends upon the geometry of its path through the counter. The straight portion, or plateau, between V_t and V_b, the breakdown voltage, is the Geiger region. The finite slope of the plateau is due to the extension of the sensitive volume of the counter and to an increase in the probability for the production of spurious counts, as the anode voltage is increased. The breakdown voltage, V_b, occurs at the end of the Geiger region, where spurious pulses occur due to failure of the quenching mechanism leading, at still higher voltages, to a continuous discharge. The operating voltage of the counter is arbitrarily chosen as $V_t + 50$ V or the voltage halfway along the plateau, whichever is the lower.

The counting rate from a point source at a distance d from the counter window will obviously depend upon the solid angle subtended by the window at the source. For a counter window of radius r this angle is given by

$$\Omega = 2\pi(1 - \cos \alpha)$$

Experiment 4.1

where α is the semi-angle of the cone subtended by the window at the source, and the geometrical efficiency is given by

$$\frac{\Omega}{4\pi} = \tfrac{1}{2}(1 - \cos \alpha) = \frac{1}{2}\left[1 - \frac{d}{\sqrt{(d^2 + r^2)}}\right]$$

The relative efficiency of the counter for a source placed at various distances, d, from the counter window can be calculated using this formula. For this calculation it must be assumed that the source is a point source, that the sensitive volume of the counter is situated

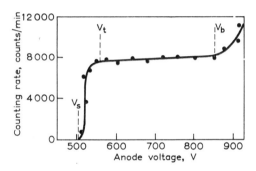

Fig E.4.1 Typical "plateau" for a Geiger–Müller counter

just behind the counter window and extends outwards to the full diameter of the counter, and that there is no absorption of the radiations by the air space between the source and the window or by the window itself. Since these assumptions are not generally valid, it is better to determine the efficiency of the counter at different source geometries experimentally, by counting a standard source at various distances from the counter window. The efficiency of a Geiger–Müller counter is dependent upon both the operating voltage and the age of the counter, so that it is advisable to make routine checks on its efficiency by counting a standard source at frequent intervals.

Experimental

1. Place a radioactive source in a Geiger–Müller counter castle.
2. Slowly raise the anode voltage until the scaler just begins to count. Take a minute count of the source. Increase the voltage by 5 V steps between V_s and V_t, taking a minute count at each voltage. Above V_t increase the voltage by 50 V steps, taking a count at each voltage. Do not raise the anode voltage much above the breakdown value; otherwise the counter may be damaged, particularly if it is of the organic quenched type.

3. Plot the counting rate against the anode voltage for each result as it is obtained, starting the counting rate scale (the ordinate) from zero.
4. Record: (i) the starting voltage, (ii) the threshold voltage, (iii) the plateau slope, and (iv) the normal operating voltage. The *plateau slope* is defined as the percentage change in counting rate for a 1 V change in anode voltage.
5. Determine the counting rate for an absolutely standardized radioactive source* in the various shelves of the counter castle. Calculate the percentage efficiency of the counter for each of these positions from the known disintegration rate of the source and the observed counting rates. Compare the relative efficiencies with those calculated using the equation above.

Further work

Determine and compare the efficiencies of the counter for different source geometries for an α-particle source and for several β-particle sources of different energies.

EXPERIMENT 4.2   DETERMINATION OF DEAD TIME OF A GEIGER–MÜLLER COUNTER

If a Geiger–Müller counter is used without a quenching circuit then the inoperative period which occurs after each count will vary somewhat from count to count. Its average value, the counter dead time, may be determined experimentally by the method described below. When a quenching circuit is used, the same methods may be used to determine the inoperative period, or paralysis time, of the quenching circuit. It has been shown in Chapter 3 that

$$n = m + nm\tau \qquad (E.4.1)$$

where n is the true counting rate for a source, m is the observed counting rate for the source, and τ the inoperative period.

Eqn. (E.4.1) may be rewritten as

$$\frac{1}{m} = \frac{1}{n} + \tau$$

which is the equation of a straight line of intercept τ on the $1/m$ axis. The dead time or paralysis time, τ, may therefore be determined from a plot of $1/m$ against $1/n$ for a series of sources of known relative activity. The quantities $1/n$, are not known absolutely since the true

---

* Obtainable from the Radiochemical Centre. Alternatively a weighed sample of uranium oxide may be used, the source being covered with a thin aluminium foil to absorb the α-particles emitted. Natural U_3O_8 has a β-disintegration rate due to ^{234m}Pa of 622 disintegrations per minute per milligram.

Experiment 4.2

counting rate, n, cannot be directly determined. It is, however, only necessary to plot quantities proportional to $1/n$ and these may be calculated from the known relative activities of the sources.

The method can only be applied if the scaler can function at counting rates up to at least 30 000 counts/min.

Experimental

Labelled material   Phosphorus-32-labelled sodium orthophosphate (Radiochemical Centre catalogue number PBS 1)

Recommended activity   1·0 μCi

1. To 1 μCi of a carrier-free phosphorus-32 solution add about 5 mg of sodium phosphate carrier and make the total volume up to about 10 cm^3 with distilled water.
2. Accurately dispense the following volumes of this solution into 1 cm^3 aluminium dimple counting trays, filling only the dimple. The last two solutions may have to be added to the counting tray in two aliquots, evaporating between additions, so that a constant source geometry is preserved as nearly as possible. The orders of magnitude of the counting rates to be expected close to the window of an end-window Geiger–Müller counter are given below

cm^3	counts/min
0·10	10 000
0·15	15 000
0·20	20 000
0·25	25 000
0·30	30 000

    Evaporate each of the solutions to dryness under an infra-red lamp.
3. Determine the counting rate of each of these sources under the same conditions of geometry, placing them on the top shelf in a Geiger–Müller counter castle. The counting time should in each case be long enough to ensure that the 95% probable random error is not greater than 1% (see Experiment 5).
4. Plot $1/m$ against $1/v$, where v is the volume of the solution used, and determine the dead time or paralysis time from the intercept on the $1/m$ axis. The $1/m$ scale should be selected to give the precision necessary to determine the intercept, τ.

Further work

The method may also be used to determine the dead time of a liquid-sample Geiger–Müller counter, the volumes of phosphorus-32 solution being dispensed directly into the counter and made up to about 10 cm^3 in each case.

Experiment 5
The statistics of radioactive decay

The radioactive decay law in its simplest form gives the rate of decay of a radioisotope preparation containing N atoms as

$$\frac{dN}{dt} = -\lambda N$$

where λ is the radioactive decay constant for the radioisotope. The number of atoms in the preparation, N, is clearly a precise quantity which is related to the mass of the radioisotope present. The radioactive decay constant, λ, is also a quantity which can be precisely evaluated. However, measurements of radioactivity are subject to statistical variations. If successive counts of equal duration are made on an essentially constant radioactive source (i.e. a source whose half-life is very long compared to the duration of the experiment), the number of disintegrations recorded in each time interval will not be equal, even though all the possible errors in measurement are absent.

The explanation of such variations lies in the fact that λ is the probability of decay of a single atom and there is therefore a probability of $1 - \lambda$ that the atom will not decay in that particular time interval. Thus for a large number of atoms, N, the quantity λN does not imply that we shall always observe a decay rate of precisely λN but only that the average decay rate, calculated from the results of a large number of measurements, will be equal to λN.

The situation can be made clearer by the analogy to the tossing of coins. Suppose that we toss 16 coins simultaneously and then count the number that have fallen "heads". We will regard this as the analogue of the number of atoms that have decayed. Coins which come down "tails" will be analogous to those atoms that did not decay and are therefore not observable, so we will ignore them. The probability, λ, that a coin will fall "heads" is 0·5, and the number which on average will fall "heads" in a throw will be λN, or in this case 8. However, from experience we know that we shall not observe 8 "heads" in every throw. Experimentally from a large number of throws we can determine the frequency or probability of observing any particular number of "heads" from 1 to 16 in a

Experiment 5 125

Table 14 The probabilities of observing n "heads" in a throw of 16 coins

n	$P(n)$
8	0·14
7, 9	0·12
6, 10	0·10
5, 11	0·077
4, 12	0·052
3, 13	0·015
2, 14	0·007
1, 15	0·003

throw, or alternatively this may be calculated from the Gaussian distribution (eqn. (E.5.2)). These probabilities are given in Table 14 and are represented graphically by plotting the probability $P(n)$ of observing n "heads" in a throw against n. This has been plotted in Fig. E.5.1 for the case in which the true average number of "heads", N, is 8, that is for throws of 16 coins. Such a graph or distribution gives the probability of observing any particular number of "heads" in a throw. We can also draw horizontal lines on the graph to denote the limits of the deviations from the average within which any percentage of the total number of throws will lie. For example, in this case it is found that 73 % of all throws have between 5 and 11 "heads".

We now return to the problem of random or statistical variation in observations of the counting rate of a radioactive source. In a radioisotope source with a long half-life, the number of atoms, N, will not decrease appreciably within the time required to make our

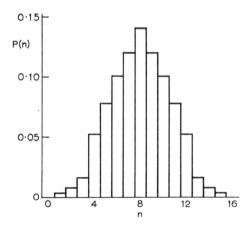

Fig E.5.1 The variation of the probability, $P(n)$, of observing n "heads" in a throw of 16 coins, with n

observations of counting rate. In other words, the number of radioactive atoms which decay is small compared to the total number of atoms present, so that on a statistical basis the decay of an atom is a rare event. The chance that any one atom will decay is therefore small compared to the chance that it will not decay, and the probability, $P(n)$, of n such disintegrations occurring in a given time interval follows the Poisson distribution law:

$$P(n) = \frac{\bar{N}^{-n}}{n!} e^{-\bar{N}} \tag{E.5.1}$$

where $\bar{N}$ is the average number of disintegrations in the same time interval, determined from a large number of observations. The form

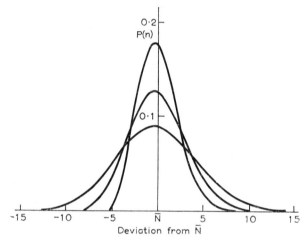

Fig E.5.2 Graphs of the Poisson distribution
Showing the probabilities, $P(n)$, of positive and negative deviations from the true average $\bar{N}$, for values of $\bar{N}$ of 5, 10 and 20

of the Poisson distribution for various values of $\bar{N}$ is illustrated in Fig. E.5.2. It is a feature of this distribution that, when the number of disintegrations recorded, n, is small, the probability of this number being slightly greater than the average, $\bar{N}$, is not the same as the probability of its being the same amount less than the average. This is indicated by the asymmetrical nature of the curves. As $\bar{N}$ increases, the Poisson distribution becomes more symmetrical with respect to the average and approaches a Gaussian distribution, for which

$$P(n) = \frac{1}{\sqrt{(2\pi\bar{N})}} e^{-(\bar{N}-n)^2/2\bar{N}} \tag{E.5.2}$$

Experiment 5   127

For a set of measurements of equal duration on a source, the standard deviation, σ, is defined as the root mean square of all the deviations from the average. It is a property of the Poisson distribution that σ is equal to the square root of the average count, $\bar{N}$. In practice it is therefore not necessary to make a large number of counts upon a source to determine σ, since the following approximations can be made:

1. $\sigma \approx \sqrt{\bar{n}}$, where $\bar{n}$ is the average count calculated from a small number of counts.
2. $\sigma \approx \sqrt{n}$, where n is a single count.

The value of the average, $\bar{n}$, when determined from a small number of measurements of n, is generally different from the true average, $\bar{N}$, which must be determined from a large number of measurements of

Table 15 Variation of error in counting rate with number of counts recorded

Time to record n counts with		Counts recorded, n	Standard deviation, σ	95·45% probable percentage error in n
A source of 10 000 cpm	A source of 50 cpm			
50 min	167 h	500 000	707	±0·3
10 min	34 h	100 000	316	±0·6
5 min	17 h	50 000	223	±0·9
1 min	3 h	10 000	100	±2·0
1 min	40 min	2 000	45	±4·5
1 min	2 min	100	10	±10·0

n. It can be shown that, for a Gaussian distribution, 68·27% of all measurements of n lie within $\pm 1\sigma$ of $\bar{N}$, and that 95·45% lie within $\pm 2\sigma$ of $\bar{N}$. If the above approximations are made, therefore, for a single count, n, on a radioactive source there is a 68·27% chance that it lies within $\pm 1\sigma$, or $\pm\sqrt{n}$, of the true average count, $\bar{N}$, and a 95·45% chance that it lies within $\pm 2\sigma$, or $\pm 2\sqrt{n}$, of the true average. These calculations apply to a Gaussian distribution and in practice apply to most radioactivity determinations, provided that more than 100 disintegrations are observed for each count, n, since the Poisson distribution approaches a Gaussian distribution for $\bar{N} > 100$.

For a single measurement in which a count of n radioactive disintegrations is taken in a time t, the 95·45% probable error in the count, n, is $\pm 2\sqrt{n}$ approximately, and the 95·45% probable error in the counting rate n/t, is $\pm 2\sqrt{n}/t$ approximately. The percentage errors in count and counting rate are identical. Table 15 gives the 95·45% probable errors in count when various total counts are recorded, and also gives the times required to record these counts for two sources of different activities, one a typical tracer level source of 10 000 counts per minute and one of 50 counts per minute, i.e. not much above background level. Inspection of this table reveals that the statistical error in a count on a given source may be decreased by increasing the counting time. In most tracer level experiments it is reasonable to aim at a probable random error of not more than $\pm 2\%$, i.e. to record 10 000 counts. The activity from sources of low activity or from the natural background cannot generally be measured with this order of accuracy. In most experimental work it is necessary to determine the counting rates of more than one source and to combine them arithmetically. The standard deviations of various combined quantities are related to the standard deviations of the individual errors and are given in Table 16.

Table 16 The combination of errors in arithmetic operations

Quantity	Standard deviation
n	$\sigma_n = \sqrt{n}$
r	$\sigma_r = \sqrt{n/t}$
$r_1 \pm r_2$	$(\sigma_{r_1}^2 + \sigma_{r_2}^2)^{1/2}$
$r_1 \times r_2$	$(r_2^2 \sigma_{r_1}^2 + r_1^2 \sigma_{r_2}^2)^{1/2}$
$\dfrac{r_1}{r_2}$	$\dfrac{r_1}{r_2}(r_1^2 \sigma_{r_1}^2 + r_2^2 \sigma_{r_2}^2)^{1/2}$

For n counts recorded in time t, the counting rate, r, is n/t.

THE OPTIMUM DIVISION OF COUNTING TIME BETWEEN SOURCE AND BACKGROUND

It is frequently necessary to determine the counting rate of a source, or of each of a series of similar sources, and then to correct these counting rates for the contribution made by the natural background. The random error of the corrected counting rates may be minimized by the correct division of the counting time available between the sources and the background. If r_b is the counting rate due to background and r_s the counting rate of each of x similar sources then the optimum division of the available counting time is given by

$$\frac{t_b}{t_s} = \sqrt{x \frac{r_b}{r_s}}$$

where t_b is the counting time for the background and t_s the counting time for each of the sources. The optimum counting times may be calculated from this expression after approximate counting rates for the source and background have been determined using short counting times. Since r_s, the counting rate due to the source plus background, must always be greater than r_b, the counting rate due to background, it is evident that, even when only one source is to be counted and corrected for background, t_b cannot be greater than t_s, i.e. it is never worth counting the background for longer than the source. In many cases the counting rate due to background is very small compared to that due to the source, and it may even be negligible compared to the statistical error in the counting rate of the source itself. For most tracer level work, therefore, the background need not be determined with great accuracy. In low-level counting, where the counting rate due to the source plus background may be only marginally greater than that of the background alone, the optimum division of the counting time between the source and the background becomes important if reliable results are to be obtained.

THE CHI-SQUARED TEST FOR NON-RANDOM ERRORS

In the determination of the activity of a radioactive sample, as well as the errors due to the random nature of radioactive decay, errors of a non-random nature can also arise. These may be due to faults in the counting equipment, they may be personal errors in operating the equipment or in recording results, and some errors in the preparation of samples for counting. The presence of such errors may be revealed by making a statistical analysis or test on a series of repeated measurements, in the case of the first two types of error on the same source, and in the case of the third type on a series of similar sources.

Whereas the statistical errors will follow a Gaussian distribution, errors of a non-random nature will not, and if a series of measurements cannot be fitted to a Gaussian distribution curve, then non-random errors must be present. The chi-squared test is a method of testing the goodness of fit of a series of observations to the Gaussian distribution. If non-random errors are absent, then the values of χ^2 as defined below should lie between the limits quoted in Table 17 for various numbers of observations:

$$\chi^2 = \frac{\sum_{i=1}^{i=q}(\bar{n} - n_i)^2}{\bar{n}}$$

where $\bar{n}$ is the average count observed, n_i is the number of counts in the lth observation, and q is the number of observations. If a series of observations fits a Gaussian distribution then there is a 95% probability that χ^2 will be greater than or equal to the lower limit quoted in Table 17, but only a 5% probability that χ^2 will be greater than or equal to the upper limit quoted. Thus, for a series of 10

Table 17 The limits of the quantity χ^2 for sets of counts with random errors

Number of observations	Lower limit for χ^2	Upper limit for χ^2
3	0·103	5·99
4	0·352	7·81
5	0·711	9·49
6	1·14	11·07
7	1·63	12·59
8	2·17	14·07
9	2·73	15·51
10	3·33	16·92
15	6·57	23·68
20	10·12	30·14
25	13·85	36·42
30	17·71	42·56

observations, if χ^2 lies outside the limits 3·33 to 16·92, it is very probable that errors of a non-random nature are present.

In applying the chi-squared test the number of counts recorded in a single counting observation should be large enough to make the statistical error less than the accuracy required for the activity determination. Thus if 10 000 counts are recorded for each observation and for a series of observations χ^2 lies between the expected limits, it can be concluded that non-random errors of magnitude greater than about $\pm 2\%$ are absent.

Experimental

1. Place a radioactive source in the castle of a Geiger–Müller counter in such a position that it gives a counting rate of about 1000 counts per minute. Record at least 30 one-minute counts of that source. The value of 1000 counts/min has been selected to emphasize that in a one-minute count the 95% probable error is about 6%. A more accurate determination of counting rate from a small number of counting observations would require a longer counting time.

2. Calculate the average count, the deviation of each count from the average and the standard deviation, σ, for the set of counts.

3. Determine the percentage of the counts which lie within $\pm 2\sqrt{\bar{n}}$ and $\pm 2\sigma$ of the average.

4. Compare the standard deviation with the square root of the average count.

5. Alter the position of the source in the castle or replace it with a more active source so that a counting rate of about 10 000 counts per minute is obtained. Record 6 one-minute counts of the source, making a deliberate timing error of 5 s in the last of them. Apply the chi-squared test to the first five and the last five of the counts. Comment on your results.

Experiment 6

Determination of the energy of alpha particles by measurement of their range in air

The α-particle is a helium nucleus which is emitted from a radioactive nucleus. α-particles are emitted with one or more discrete energies depending on the number of energy states in which the daughter nuclei may be left. In passing through matter the α-particle, like the β-particle, loses energy by excitation or ionization of the atoms or molecules of the absorber. The number of ions produced in each unit length of path (specific ionization) is primarily dependent on the velocity of the particle, and its value increases as the velocity decreases. This is to be expected as the α-particle spends an increasing length of time in the vicinity of the bound electrons of the absorber atom, resulting in a higher probability of their being excited or ionized. Due to the larger mass of the α-particle its velocity will be less than that of a β-particle of the same kinetic energy, and therefore its specific ionization will be greater and its total path in matter much shorter. A 3 MeV α-particle has a range in air of about 2 cm compared with about 1100 cm for a β-particle of the same energy.

The α-particles, because of their relatively large mass, are not deflected by interactions with the electrons of the absorber. Their paths are therefore straight lines, a fact which can readily be demonstrated using a cloud chamber (see Experiment 13). For a collimated beam of monoenergetic α-particles from a thin radioactive source the number of particles reaching a detector per unit time is plotted against the thickness of absorber between the source and the detector in Fig. E.6.1. The variation in the ranges of monoenergetic α-particles arises because of the statistical nature of the processes by which the particles lose their energies. Thus, there are fluctuations in both the energy lost in each collision with an electron and in the number of collisions per centimetre of path. The distribution of ranges about the mean range is indicated by the broken curve in Fig. E.6.1. The mean range, R_m, is defined as the point at which the number of α-particles reaching the detector per unit time falls to one-half of the initial value. The extrapolated range, R_e, is obtained by a straight-line extrapolation from the point on the curve determined by R_m. The difference between the mean range and the extrapolated range is known as the "straggling" and amounts to about 1 % of the range

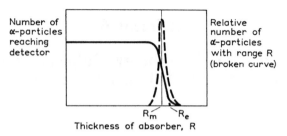

Fig E.6.1 Absorption of α-particles

for 5 MeV α-particles. If the α-particles are not well collimated, as is likely to be the case in the conditions used for the experiments described below, the range plot will show a negative slope and the statistical straggling will be masked by the greater effect of the poor collimation.

ALPHA-PARTICLE RANGE/ENERGY RELATIONSHIPS

Empirical relationships have been developed between the mean range of α-particles in air and their energy. Fig. E.6.2 is a graphical representation of these relationships. Table 18 gives the energies and the ranges in air of the α-particles emitted by some frequently used

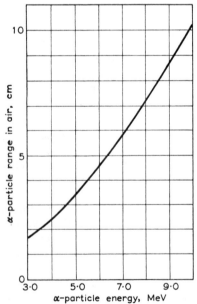

Fig E.6.2 Empirical plot of the range of α-particles in air (at s.t.p.) against α-particle energy

Experiment 6.1 133

Table 18 Energies and corresponding ranges in air of the α-particles from some α-emitting sources

Source	α-energy, MeV	Range in air, cm	Half-life
Radium source			
^{226}Ra	4·78	3·45	1620 y
	4·59 (6%)	3·0	
^{222}Rn	5·48	3·9	3·825 d
^{218}Po	6·00	4·65	3·05 m
^{214}Po	7·68	6·8	$1·6 \times 10^{-4}$ s
^{210}Po	5·3	3·8	
^{212}Pb/^{212}Bi source from thorium cow			
^{212}Po	8·78	8·7	3×10^{-7} s
Americium source			
^{241}Am	5·48	4·0	458 y
Plutonium source			
^{239}Pu	5·15	3·6	$2·4 \times 10^{4}$ y

sources. The range in aluminium, R_{Al}, may be calculated from the relation

$$R_{Al} \text{ (mg/cm}^2\text{)} = 1·57 \, R_m \text{ (cm)}$$

where R_m is the mean range of the α-particles in air at standard temperature and pressure.

EXPERIMENT 6.1 MEASUREMENT OF ALPHA-PARTICLE RANGE USING A GEIGER–MÜLLER COUNTER OR A SOLID-STATE DETECTOR

Experimental

1. Arrange a Geiger–Müller counter or a solid-state detector with its window in the vertical plane, and place a collimated monoenergetic α-source directly opposite and close to the counter window, being very careful not to let the collimator touch the fragile counter window. A cylinder of aluminium foil about 1 cm long and 0·5 cm in diameter fixed to the source forms a suitable collimator. The source should be mounted so that it can be moved along a track with a vernier scale.

2. Slowly move the source away from the counter, and without actually measuring the count rates, determine the approximate distance at which the count rate falls to background.

3. Measure the counting rate as a function of the source-to-counter distance. Move the source towards the counter in steps of about 0·25 mm over the region of straggling where the count rate increases rapidly. Increase the steps to about 3 to 5 mm when the counting rate shows that the plateau of Fig. E.6.1 has been reached.

The counting times in each case should be a compromise between good statistical accuracy and the time available for the experiment.

4. Correct the counting rates for counter dead time in the case of a Geiger–Müller counter and for counter background. The source-to-counter distances must also be corrected for absorption within the Geiger–Müller counter window. Each 1·4 mg/cm^2 of mica window is equivalent to an air distance of 1 cm for α-particle absorption. The very thin evaporated gold window of a solid-state detector has a negligible effect on the absorption of the α-particles.

5. Plot the corrected counting rate against corrected air thickness.

6. Determine the mean and extrapolated ranges of the α-particles. Estimate the energy of the α-particles using Fig. E.6.2, making the approximation that the mean and extrapolated ranges are the same. As has been explained above, it is not possible to investigate straggling under the poor collimation conditions used in this experiment.

Further work

If the apparatus is placed in a tube through which hydrogen can be passed the range of the α-particles in hydrogen may be determined.

EXPERIMENT 6.2 MEASUREMENT OF ALPHA-PARTICLE RANGE USING A PULSE ELECTROSCOPE

Experimental

1. Set up the pulse electroscope with an α-particle source mounted on the central electrode and place an ionization chamber of adjustable height in position above the electroscope (see Fig. E.3.2). Apply a potential sufficient to saturate the ion chamber through a safety resistor of about 10 MΩ.

2. Determine the discharge frequency of the electroscope with the top of the ionization chamber at various distances from the source. Start with the top of the ionization chamber as close as possible to the source (but not touching), and move it away in about 0·25 cm steps until the discharge frequency reaches a saturation value.

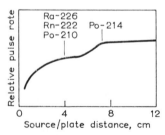

Fig E.6.3 Determination of the α-particle ranges due to a radium source with a pulse electroscope

Experiment 6.3 135

3. Plot "discharge frequency" against distance of the top of the ionization chamber from the source. A curve such as that in Fig. E.6.3 will be obtained.

4. Find the ranges in air of the radium-226 and polonium-214 groups of α-particles as indicated in Fig. E.6.3. Using these ranges determine the α-particle energies from Fig. E.6.2.

Note. The discharge frequency is proportional to the saturation ionization current in the chamber. When the top of the ionization chamber is very close to the source only a small volume of air is ionized by the α-particles. The ions so formed give rise to a small saturation current. As the distance d is increased the volume of air which the α-particles ionize becomes greater and so the saturation current increases. The saturation current reaches a maximum when d is greater than the range of the α-particles in air, since the number of ions formed and collected is then constant. When the plate is close to the α-particle source, those particles which collide with the plate dissipate a part of their kinetic energy as thermal energy within the plate.

EXPERIMENT 6.3 ABSORPTION OF ALPHA-PARTICLES IN ALUMINIUM FOILS

Experimental

(a) Using Geiger–Müller counter

1. Place an α-particle source about 1 cm from the window of an end-window Geiger–Müller counter. Determine the counting rate with good statistical accuracy.

2. Cover the source with thin aluminium foil 0·005 mm thick and determine the counting rate again. Repeat, adding successive foils of the same thickness until the counting rate is approximately constant.

3. Correct the aluminium foil thicknesses for the counter window thickness and the air space between the source and the counter window, expressing the corrected foil thickness in terms of mg/cm² of aluminium:

1 foil of thickness 0·005 mm $\equiv$ 1·35 mg/cm²
1 cm of air $\equiv$ approximately 1·7 mg/cm² of aluminium
1 mg/cm² of mica $\equiv$ approximately 1·2 mg/cm² of aluminium

Correct the counting rates for paralysis time.

4. Plot $\log_{10}$ (corrected counting rate) against the corrected aluminium absorber thickness in mg/cm².

5. Estimate from the graph the range in mg/cm² of aluminium. Use the relationship

$$R_{Al}(\text{mg/cm}^2) = 1\cdot 57 R_m(\text{cm})$$

where R_{Al} is the range in aluminium and R_m the range in air at standard temperature and pressure, to estimate the corresponding range in centimetres of air. Compare this range in air with that obtained directly in Experiment 6.1.

(b) *Using pulse electroscope*

1. Arrange the pulse electroscope and an α-particle source as described for Experiment 6.2.
2. Cover the source with aluminium foil of thickness 0·005 mm and determine the discharge frequency again. Repeat, adding successive foils of the same thickness until the discharge frequency is approximately constant. Take care not to include any air space between the foils.
3. Express the foil thicknesses in terms of mg/cm² of aluminium (1 foil of thickness 0·005 mm = 1·35 mg/cm²).
4. Plot $\log_{10}$ (discharge frequency) against aluminium absorber thickness in mg/cm².
5. Estimate the range in mg/cm² of aluminium from the graph. Use the relationship

$$R_{Al}(\text{gm/cm}^2) = 1\cdot 57 R_m(\text{cm})$$

where R_{Al} is ther ange in aluminium and R_m the range in air at standard temperature and pressure, to estimate the corresponding range in centimetres of air. Compare this range in air with that obtained directly in Experiment 6.1.

(c) *Using quartz-fibre electroscope dosimeter*

In this experiment the ionization current in an ion chamber is measured by a quartz-fibre electroscope (dosimeter) mounted on top of the ion chamber (Fig. E.6.4).

1. Place the α-source in the bottom of the ionization chamber, charge the electroscope, and measure the time required for it to discharge between fixed points on the scale.
2. Cover the source with successive thicknesses of aluminium foil as in the previous experiments and again charge the electroscope and measure the time required for discharge.
3. Plot $\log_{10}$ (discharge rate) against the aluminium absorber thickness in mg/cm².
4. Determine the range and energy of the α-particles as before.

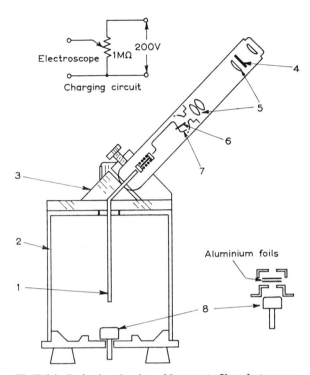

Fig E.6.4 Ionization chamber with a quartz fibre electroscope

Insets: an exploded view of a device for holding aluminium foil absorbers, and electroscope charging circuit

1. Positive electrode
2. Earthed electrode
3. Perspex prism
4. Graticule
5. Optical system
6. Quartz fibre
7. Fibre support and electrode
8. Source

Note. In all these methods it is important to ensure that no air is trapped between successive aluminium foils as this would also contribute to the absorption of the α-particle energy and thus effectively increase the thickness of absorbing material. This condition can be achieved by the use of a device to hold the foils in close contact, as shown in Fig. E.6.4.

EXPERIMENT 6.4 DETERMINATION OF ALPHA-PARTICLE RANGE IN AIR USING A SPARK COUNTER

The spark counter consists of a grid of fine wires separated from a metal plate by a small air gap (Fig. E.6.5). The wires are held at a high potential, about 6 kV, with respect to the plate. If a particle of high specific ionization passes close to one of the wires, it can initiate

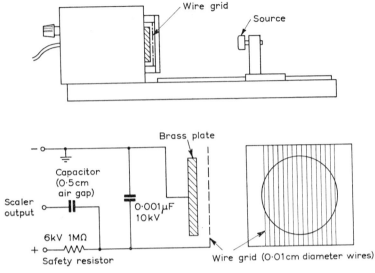

Fig E.6.5 Diagram of a spark counter

a discharge in the region of high field around the wire and a visible spark results. The device provides a very sensitive means of estimating the range of α-particles in air. The range of the most energetic α-particle emitted by a source can be accurately determined by finding the minimum distance between the wire grid and the source at which sparking just fails to occur.

If the spark counter is connected to a scaler, the sparking rate may be investigated as a function of the distance between the grid and the source. Since the specific ionization of an α-particle increases towards the end of its range, the probability of causing a spark will increase at the end of the range. This will result in a peak in the sparking rate, as shown in Fig. E.6.6.

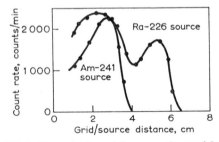

Fig E.6.6 Determination of the range of α-particles in air using a spark counter

Experiment 6.4 139

Experimental

1. Connect the spark counter to a 6 kV d.c. power supply through a 1 MΩ safety resistor. With no radioactive source present, adjust the voltage applied to the counter until sparks are observed due to electrical breakdown of the air between the wires and the plate. Then reduce the voltage until such breakdown just ceases.

2. Place an α-emitting radioactive source on the track of the counter close to the grid. Gradually move the source away from the grid until sparking just ceases. The distance between the source and the grid at which sparking ceases is equal to the range in air of the α-particles. In this manner visually determine the ranges of the α-particles from a series of α-emitting sources.

3. Connect the spark counter to a scaler and investigate the variation of sparking rate with the grid-to-source distance for a radium and an americium source.

4. From their measured ranges in air determine the energies of the α-particles using Fig. E.6.2.

Experiment 7

The determination of the range and energy of beta particles

Electrons, both positive and negative, may be emitted spontaneously from a nucleus in the process of β-decay. The energies of such β-particles vary from 18 keV for tritium (hydrogen-3), for example, to 3·6 MeV for potassium-42. Nuclear β-particles commonly have velocities which are between 90% and 99% of the speed of light, and the mass of the β-particle, at such velocities, must be replaced by the relativistic mass, m:

$$m = m_0(1 - v^2/c^2)^{-1/2}$$

where v is the particle velocity, c is the velocity of light and m_0 is the rest mass of the electron. The classical kinetic energy equation,

$$E = m_0 v^2$$

is a good approximation only for β-particles of energies less than 100 eV. For higher energies, the kinetic energy is correctly given by the relativistic equation:

$$E = (m - m_0)c^2$$

where m is the relativistic mass defined above. Table 19 lists the kinetic energies of β-particles with various velocities.

Table 19 Variation of the velocity (v) and mass (m) of β-particles with their kinetic energy (E)

v/c	m/m_0	E (MeV)
0·01	1·0005	0·00027
0·1	1·0073	0·0037
0·5	1·1547	0·0774
0·7	1·3843	0·192
0·8	1·6667	0·334
0·9	2·2942	0·647
0·95	3·2023	1·101
0·99	7·092	3·046
0·999	22·47	10·74

c = Velocity of light = $2·998 \times 10^8$ m/s
m_0 = Rest mass of electron = $9·108 \times 10^{-31}$ kg

An analysis of the numbers of β-particles of various energies which are emitted by a given radionuclide results in a spectrum such as that in Fig. 1.5. Electrons are emitted from the nucleus with a continuous range of energies from zero up to a maximum energy, E_{max}. Pauli suggested in 1927 that, in order to make conservation of energy possible for the β-disintegration process, another particle, which he called the *antineutrino*, is emitted simultaneously with the β-particle, and that the energy of the disintegration is shared between the two particles. The antineutrino carries no charge and has a negligible rest mass. The existence of this particle also makes possible the conservation of angular momentum due to particle spins, which is possible for the four-particle event:

$$_Z^A X \rightarrow {_{Z+1}^A}Y + \beta^- + \bar{\nu} \text{ (antineutrino)}$$

but not for the three-particle event:

$$_Z^A X \rightarrow {_{Z+1}^A}Y + \beta^-$$

The antineutrino may acquire any percentage, from 0 to 100% of the disintegration energy.

On passing through matter, β-particles lose their energy by collisions with the extra-nuclear electrons of atoms, causing excitation or ionization of the atoms of the material in which they are absorbed. The number of electron interactions which occur in unit path length of the β-particles is proportional to the electron density of the absorbing material. It can readily be shown that, if the thickness of an absorber is expressed in terms of its mass per unit area, then this *thickness* is approximately proportional to the number of electrons per unit area of the absorber for absorbers of low atomic number. Provided that the absorber thickness is expressed in terms of mass per unit area, the range of β-particles is independent of both the atomic number and the density of the absorber, for any particular radionuclide.

Due to the continuous spectrum of β-particle energies for a given radionuclide, and also because the β-particles on collision may be scattered through large angles, there is no unique range for the β-particles from any particular source. Fig. E.7.1 is a plot of the transmitted intensity of β-particles against the thickness of aluminium absorbers for a phosphorus-32 source. The transmitted intensity falls to a constant value whose magnitude depends upon (i) any γ-ray emission which accompanies the β-disintegration (none is present for phosphorus-32), and (ii) the emission of *Bremsstrahlung*, that is, X-radiation which is emitted as the β-particles are slowed down in the electric field within the atoms of the absorber. The latter effect increases in importance for absorbers of higher atomic number.

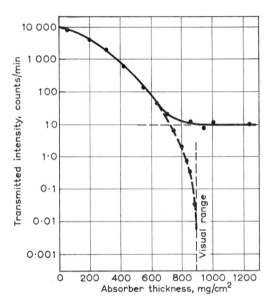

Fig E.7.1 Absorption of phosphorus-32 β-particles in aluminium

Various empirical relationships between the maximum range and the maximum energy of β-particles have been proposed. The following equation is due to Feather:

$$R = E - 0\cdot161$$

whereas Glendenin proposed two relationships as follows:

$$E = 1\cdot92 R^{0\cdot725} \quad \text{where } 0\cdot03 \text{ g/cm}^2 < R < 0\cdot3 \text{ g/cm}^2$$
$$E = 1\cdot85 R + 0\cdot245 \quad \text{where } R > 0\cdot3 \text{ g/cm}^2$$

In these formulae R is the maximum range of the β-particles expressed in g/cm^2, and E is the maximum energy of the particles expressed in MeV. The Glendenin relationships are given in Table 20.

The experimental determination of the maximum range of β-particles from a transmitted intensity against absorber thickness plot, such as that in Fig. E.7.1, presents some difficulty. The accuracy of the extrapolation procedure described below depends upon obtaining counting rates of good statistical accuracy at the higher values of absorber thickness and well beyond the range of the β-particles. Since counting rates will be approaching natural background in this region, this becomes more and more time consuming as the range is approached. A method which would enable the range

Experiment 7

Table 20 Range in aluminium of β-particles with kinetic energies from 0·15 MeV to 3·05 MeV (calculated from the Glendenin equations)

Energy, E, MeV	Range for energy E, g/cm²	Range for energy (E + 0·05), g/cm²
0·10	—	0·030
0·20	0·044	0·060
0·30	0·077	0·096
0·40	0·115	0·135
0·50	0·156	0·178
0·60	0·201	0·224
0·70	0·249	0·273
0·80	0·300	0·327
0·90	0·354	0·381
1·00	0·408	0·435
1·10	0·462	0·489
1·20	0·516	0·543
1·30	0·570	0·597
1·40	0·624	0·651
1·50	0·678	0·705
1·60	0·732	0·759
1·70	0·786	0·814
1·80	0·841	0·868
1·90	0·895	0·922
2·00	0·949	0·976
2·10	1·003	1·030
2·20	1·057	1·084
2·30	1·111	1·138
2·40	1·165	1·192
2·50	1·219	1·246
2·60	1·273	1·300
2·70	1·327	1·354
2·80	1·381	1·408
2·90	1·435	1·462
3·00	1·489	1·516

Other energy values may be interpolated

of the β-particles to be derived from the easily determined transmission intensities at absorber thicknesses well below the range would greatly facilitate energy measurement. Otazai and Hayashi (1965) have described the preparation of a graph paper which can be used to make a linear extrapolation to the range of the β-particles. The construction of the graph paper is based upon the theoretical similarity of shapes of the absorption curves for β-particles of various maximum energies. It enables the range and hence the energy to be determined, at least sufficiently accurately for the β-emitting isotope to be identified. The graph paper reproduced in Fig. E.7.2 was prepared by this method. The range/energy relationship incorporated is due to Otazai and Hayashi.

Experimental

1. Place the β-particle source whose energy is to be determined on the 2nd or 3rd shelf down in a Geiger–Müller counter castle.

2. Determine the counting rates due to the β-particles from the source with no absorbers present and when aluminium absorbers of various thicknesses are placed between it and the detector. It is important to place the absorbers as high as possible in the castle to minimize the entry of scattered β-particles into the detector. The counting time in each case should be chosen to give a counting rate determination of good statistical accuracy.

3. Correct the absorber thicknesses for the thickness of the counter window (when expressed in mg/cm² the counter window thickness may be added to the absorber thickness) and for the effect of the air space between the source and the counter window. (Add 1·3 mg/cm² to the absorber thickness for each centimetre of air path.)

Correct the counting rates for paralysis time losses but not for background. Counter background corrections would introduce further statistical errors in the visual range method and are insignificant for the linear extrapolation method.

4. Plot $\log_{10}$ (corrected counting rate) against the corrected absorber thickness. Draw a straight line through the points on the γ-ray or Bremsstrahlung background. This line should be parallel to the absorber thickness axis. Subtract this background from each point on the curve and plot the results again to obtain the extrapolated curve shown in Fig. E.7.1. The point at which the line asymptotic to this extrapolated curve cuts the absorber thickness axis is known as the *visual range* of the β-particles. Determine the β-particle energy from the visual range using the Feather or Glendenin equations or Table 20.

5. Plot the corrected counting rate against the absorber thickness on the linear extrapolation graph paper of E.7.2 after normalizing each point so that the corrected counting rate for zero absorber thickness is 10 000 counts/min. Draw a straight line through the points for low absorber thicknesses and extrapolate it to the β-particle range. Read the β-particle energy from the range/energy relationship.

Note. The determination of the individual counting rates due to the isotopes present in a mixture of β-emitting radioisotopes can be carried out by an absorption method in cases where the energies of

Experiment 7

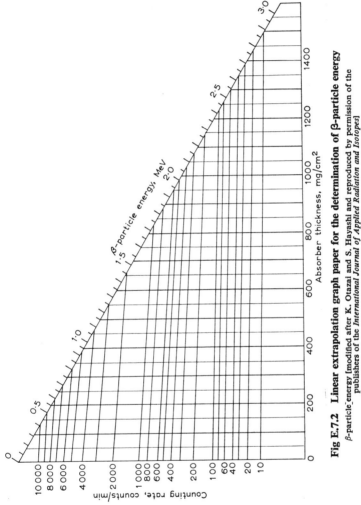

Fig E.7.2 Linear extrapolation graph paper for the determination of β-particle energy
β-particle energy [modified after K. Otazai and S. Hayashi and reproduced by permission of the publishers of the International Journal of Applied Radiation and Isotopes]

the β-particles involved are significantly different. In the study of the uptake of phosphate and sulphate ions by plants (Experiment 29.2) using phosphorus-32-labelled phosphate and sulphur-35-labelled sulphate, for example, such a method is used. It is suggested that readers who intend to perform Experiment 29.2 should investigate the absorption of phosphorus-32 and sulphur-35 β-particles as described above.

REFERENCE

OTAZAI, K. and HAYASHI, S., "A method of analysing beta-ray spectra" *International Journal of Applied Radiation and Isotopes*, **16,** p. 681 (1965)

Experiment 8

Deflection of beta particles in a magnetic field

A collimated beam of charged particles, each of velocity v, mass m and charge e, is deflected by a magnetic field of induction B into a circular path of radius r, where r is related to the other parameters by the equation:

$$Bev = mv^2/r \qquad \text{(E.8.1)}$$

or

$$rB = mv/e \qquad \text{(E.8.2)}$$

As discussed in Experiment 7, for nuclear β-particles of energy greater than 100 eV, the mass m must be replaced by the relativistic mass,

$$m = m_0(1 - v^2/c^2)^{-1/2} \qquad \text{(E.8.3)}$$

where c is the velocity of light. The *magnetic rigidity* (product of magnetic induction B and path radius r) is then given by

$$rB = \frac{m_0 v}{e}\left(1 - \frac{v^2}{c^2}\right)^{-1/2} = \frac{m_0 c}{e}\frac{v}{c}\left(1 - \frac{v^2}{c^2}\right)^{-1/2} \qquad \text{(E.8.4)}$$

The β-particles emitted in nuclear decay are not mono-energetic, as was indicated in Experiment 7. The nuclide strontium-90, for example, emits β-particles with a maximum energy of 0·54 MeV, although the majority of the β-particles emitted have an energy less than this. As the daughter radionuclide, yttrium-90, is generally in equilibrium with strontium-90, the more energetic β-particles from this nuclide are also observed. These have a maximum energy of 2·27 MeV, but most of the β-particles emitted have energies around the maximum of the yttrium-90 β-particle spectrum, say about 1·1 MeV.

Inspection of Table 19 shows that a 1·10 MeV β-particle will move with a velocity of 0·95 times the velocity of light and that its mass will be 3·2 times the rest mass of the electron. The velocity of light is $2·998 \times 10^8$ m/s, the rest mass of the electron is $9·108 \times 10^{-31}$ kg, and its charge is $1·6 \times 10^{-19}$ C. Substituting these numerical values in eqn. (E.8.4) gives the magnetic rigidity for the deflection of 0·65

MeV β-particles in magnetic fields, as follows:

$$rB = \frac{(9 \cdot 108 \times 10^{-31})(0 \cdot 95 \times 2 \cdot 998 \times 10^8)}{1 \cdot 6 \times 10^{-19} \times (1 - 0 \cdot 95^2)^{1/2}}$$
$$= 0 \cdot 00519 \text{ Wb/m}$$

The magnetic induction required to deflect the β-particles into a path of radius 0·1 m will therefore be 0·0519 T (i.e. Wb/m²). The few β-particles emitted with the maximum energy would be deflected into a path of about 0·2 m radius by such a field. A field of this order can readily be obtained with permanent magnets, and the apparatus shown in Plate 8 can be used to study the deflection of the β-particles in a magnetic field.

Experimental

1. Place a 1 μCi strontium-90 source in the collimator of the apparatus and determine the counting rate with the collimator set at various angles.

2. Plot the corrected counting rate against angle as in Fig. E.8.1. A symmetrical plot will be obtained.

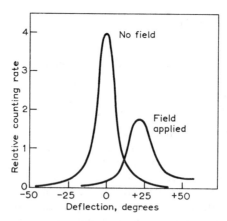

Fig E.8.1   Deflection of β-particles in a magnetic field

3. Place the permanent magnets, which should produce an induction of about 0·05 T in position as shown, and again determine the counting rate of the source at various angles.

4. Plot the corrected counting rate against angle as before. An asymmetrical plot will be obtained about a mean angle of deflection. This asymmetry is due to the nature of the β-particle spectrum.

Experiment 9
Deflection of alpha particles in a magnetic field

A collimated beam of α-particles (charge $= 2e$) will be deflected by a field of magnetic induction B into a circular path of radius r, given by

$$B(2e)v = m_\alpha v^2/r \tag{E.9.1}$$

or

$$rB = \frac{m_\alpha v}{2e} \tag{E.9.2}$$

and

$$rB = v/R \tag{E.9.3}$$

where v and m_α are the velocity and mass of the α-particle respectively, and R is the charge/mass ratio of the α-particle, the numerical value of which is 4.8235×10^7 C/kg. The α-particle velocity is related to its kinetic energy by the equation

$$E_\alpha = \tfrac{1}{2}m_\alpha v^2 \tag{E.9.4}$$

from which

$$v = \left(\frac{2E_\alpha}{m_\alpha}\right)^{1/2} \tag{E.9.5}$$

or

$$v = \left(\frac{E_\alpha(\text{MeV}) \times 10^{14}}{2.074}\right)^{1/2} \text{metres/sec} \tag{E.9.6}$$

The velocity of the 5·48 MeV α-particles emitted by americium-241 is given by eqn. (E.9.6) as 1.63×10^7 m/s. The quantity rB is now obtained from eqn. (E.9.3) as

$$rB = \frac{1.63 \times 10^7}{4.823 \times 10^7} = 0.338 \text{ Wb/m}$$

A high-field permanent magnet (such as the Eclipse Major) will have a magnetic induction of about 0·16 T (i.e. Wb/m²), and

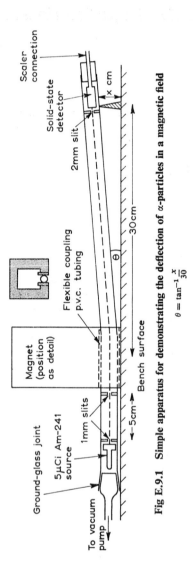

Fig E.9.1 Simple apparatus for demonstrating the deflection of α-particles in a magnetic field

$$\theta = \tan^{-1}\frac{x}{30}$$

Experiment 9 151

americium-241 α-particles will therefore be deflected by it into a path of radius 2·1 m.

Experimentally, it is therefore very much more difficult to demonstrate the deflection of α-particles in a magnetic field than that of β-particles. The high value of the magnetic rigidity, rB, means that, even with the strongest permanent magnets, α-particles can only be deflected into paths of large radius. The angular dispersion observed when they pass through fields of limited dimensions is therefore small.

It is, nevertheless, just possible to demonstrate α-particle deflection in a magnetic field using the relatively simple apparatus shown in Fig. E.9.1. It is necessary to evacuate the apparatus to increase the path length of the α-particles. A collimated beam of α-particles is deflected by the field of a permanent magnet and detected by a solid-state detector. This detector has the ability to operate in a vacuum and also has the advantage of small dimensions so that the small angular dispersion of the α-particles may be investigated. It can be shown from geometrical considerations that, for americium-241 α-particles passing through a field of induction of 0·16 T for a distance of 5 cm, the lateral displacement of the α-particle beam 30 cm from the centre of the field will be about 7 mm. It is necessary to design the collimator system so that the collimator dispersion is less that the magnetic-field dispersion of the α-particles. The counting rate to be expected from a 5 μCi source at a distance of 30 cm in a vacuum, for a small solid-state detector, is of the order of 200 counts/minute.

Experimental

1. Construct an apparatus based on the dimensions given in Fig. E.9.1 which will enable the angular dispersion of the α-particles to be measured. The slits may be made from paper or thin cardboard.

2. Place a 5 μCi americium-241 source in the collimator and evacuate the apparatus to better than 130 Pa (1 mm Hg).

3. Investigate the α-particle count rate as the detector is moved out of line with the collimator. Calculate the angle of the detector tube to the incident α-particle direction by measuring the height to which the end of the tube is raised above the bench surface.

4. Place a permanent magnet with an induction of about 0·16 T (e.g. an Eclipse Major magnet) so that the α-particle path in the region of the flexible connector passes between its poles.

5. Again investigate the α-particle count rate as the detector is angularly moved away from the collimator direction.

Experiment 10

Scattering of alpha particles by the nucleus of the atom

Rutherford introduced his nuclear theory of the atom to explain the experimentally observed large angle scattering of α-particles. (Rutherford, E., *Philosophical Magazine* 1911, pp. 669–88). He suggested that the atom consisted of a positively charged nucleus of very small dimensions surrounded by a number of extra-nuclear electrons to preserve electrical neutrality. Almost the entire mass of the atom resided in the nucleus, and those α-particles which passed close to the nucleus would undergo Coulomb repulsion forces and

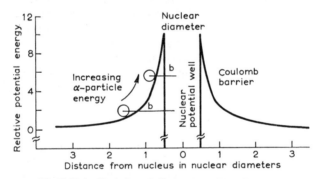

Fig E.10.1 Variation of Coulomb repulsion forces with distance from the nucleus

Showing the variation of b with α-particle energy for α-particle scattering

would thus be scattered. The angle of scattering would depend upon the distance of closest approach to the nucleus. An α-particle which approaches a nucleus of charge Ze will experience a repulsive force F, given by

$$F = \frac{2Ze^2}{4\pi\epsilon_0 r^2} \text{ newtons} \quad (E.10.1)$$

where r is the distance between the α-particle and the nucleus and ϵ_0 is the permittivity of a vacuum. Rutherford subsequently showed that this Coulomb repulsion law held down to a very small limiting value of r, below which nuclear binding forces become operative and the α-particle may be captured by the nucleus. The nucleus may therefore be represented, as in Fig. E.10.1, by a potential energy well surrounded by a Coulomb barrier. The width of the well may be regarded as defining the diameter of the nucleus.

Rutherford showed that the Coulomb repulsion forces can cause α-particles to be deflected into a hyperbolic path as shown in Fig.

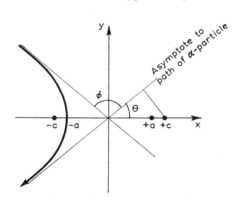

Fig E.10.2 The hyperbolic path of an α-particle in the field of a nucleus

E.10.2. The nucleus is situated at the external focus of the hyperbola, at c, the eccentricity of the hyperbola being by definition $\sec \theta = c/a$, where a is the distance from the vertex to the origin. The perpendicular distance between the nucleus and the asymptote to the incident path of the α-particle, p, is called the impact parameter. The distance between the vertex of the hyperbola and the nucleus is the distance of closest approach of the α-particle to the nucleus and is given by

$$s = c + a$$
$$= c + c \cos \theta$$
$$= p \operatorname{cosec} \theta (1 + \cos \theta) = \frac{p(1 + \cos \theta)}{\sin \theta} \quad \text{(E.10.2)}$$
$$= p \cot\left(\frac{\theta}{2}\right) \quad \text{(E.10.3)}$$

A relationship between the impact parameter, p, and the *scattering angle*, ϕ, can now be derived making use of the laws of conservation of energy and momentum. The energy of the α-particle at its point of closest approach to the nucleus will consist of a kinetic and a potential energy term and must be equal to its purely kinetic energy at a large distance from the nucleus. Thus

$$\tfrac{1}{2}mv^2 = \tfrac{1}{2}mv_0^2 + \frac{2Ze^2}{s} \tag{E.10.4}$$

where v is the velocity of the α-particle as it approaches the nucleus and v_0 is its velocity at its closest approach, distant s from the nucleus. Dividing eqn. (E.10.4) by $\tfrac{1}{2}mv^2$,

$$\frac{v_0^2}{v^2} = 1 - \frac{4Ze^2}{mv^2 s} \tag{E.10.5}$$

and substituting $b = 4Ze^2/mv^2$,

$$\frac{v_0^2}{v^2} = 1 - \frac{b}{s} \tag{E.10.6}$$

so that, finally, from eqn. (E.10.2),

$$\frac{v_0^2}{v^2} = 1 - \frac{b}{p}\frac{\sin\theta}{1+\cos\theta} \tag{E.10.7}$$

It may be noted that for a head-on collision in which the α-particle is deflected through 180°, v_0 will be zero, and to satisfy eqn. (E.10.6) b must then be equal to s, i.e. to the distance of closest approach between the α-particle and the nucleus. Since the denominator of b involves the kinetic energy of the α-particle, this distance of closest approach will become smaller as the kinetic energy of the α-particle increases. A limit is reached when the distance of closest approach becomes so small that the nuclear binding forces begin to operate and the α-particle may then be captured by the nucleus. These nuclear forces begin to operate at distances less than 10^{-13} cm, and it can be seen from the calculation below that, for an α-particle of energy about 5 MeV, the distance of closest approach to a gold nucleus is well above this limit.

Equating angular momentum of the α particle at a large distance from the nucleus to that at distance s gives

$$pmv = smv_0$$

from which

$$\frac{v_0}{v} = \frac{p}{s} = \frac{\sin\theta}{1+\cos\theta} \tag{E.10.8}$$

Substituting v_0/v from eqn. (E.10.8) in eqn. (E.10.7) yields

$$\frac{\sin^2\theta}{(1+\cos\theta)^2} = 1 - \frac{b}{p}\frac{\sin\theta}{1+\cos\theta} \tag{E.10.9}$$

Experiment 10

which on simplification gives

$$\tan \theta = \frac{2p}{b} \quad \text{(E.10.10)}$$

Since the scattering angle $\phi = \pi - 2\theta$, this equation becomes

$$\cot \frac{\phi}{2} = \frac{2p}{b} \quad \text{(E.10.11)}$$

This equation relates the scattering angle, ϕ, to the impact parameter, p, and the quantity b which involves the kinetic energy of the α-particle. For a 5 MeV α-particle undergoing scattering by a gold nucleus ($Z = 79$) b may be evaluated on substituting

$\epsilon_0 = 1/(4\pi \times 9 \times 10^9)$ F/m
$e = 1\cdot 6 \times 10^{-19}$ C

and

$mv^2 = 2 \times 5$ MeV $= 10 \times 1\cdot 6 \times 10^{-13}$ J

as follows:

$$b = \frac{4 \times 79 \times (1\cdot 6 \times 10^{-19})^2}{10 \times 1\cdot 6 \times 10^{-13}} = 4\cdot 57 \times 10^{-14}\text{ m}$$

Calculated values of the impact parameter at various scattering angles are given in Table 21 for a 5 MeV α-particle scattered by a gold nucleus. It may be noted that α-particles are deflected in the nuclear field through progressively smaller angles as the impact parameter increases. At distances of about 50 times the apparent diameter of the nucleus (which is approximately b, the distance of closest approach of the α-particle to the nucleus) the scattering angle is very small. This is, nevertheless, a very small distance compared with the diameter of the gold atom, which is about $8\cdot 3 \times 10^{-10}$ m.

Table 21 Impact parameters at various scattering angles for scattering of a 5 MeV α-particle by a gold nucleus

Scattering angle	Impact parameter, p (cm) $\times 10^{-12}$
5°	206
10°	103
30°	34
50°	19
90°	9
130°	4·2
160°	1·6
180°	0·0

The discussion so far has concerned the two-dimensional scattering of an α-particle by a single nucleus. It may be extended to scattering in all directions by a population of scattering centres or nuclei contained in a scatterer of finite dimensions. Thus the fraction, f, of α-particles which will be scattered through angles lying between ϕ_1 and ϕ_2, by a scatterer of thickness t, can be shown to be

$$f = \frac{\pi n t b^2}{4} \left[\cot^2\left(\frac{\phi_1}{2}\right) - \cot^2\left(\frac{\phi_2}{2}\right) \right] \quad \text{(E.10.12)}$$

where n is the number of nuclei per cubic centimetre of the scatterer. In the case of scattering of a 5 MeV α-particle by gold,

$$b = 4 \cdot 57 \times 10^{-12} \text{ cm}$$

and

$$n = \frac{6 \times 10^{23}}{79} \times 19 \cdot 32 = 1 \cdot 465 \times 10^{23} \text{ atoms/cm}^3$$

and the fraction of α-particles which will be scattered by angles between 160° and 180°, for a scatterer 0·000 1 cm thick, will be

$$f_{160°-180°} = \frac{\pi}{4} \times 1 \cdot 465 \times 10^2 \times 0 \cdot 0001$$

$$\times (4 \cdot 57 \times 10^{-12})^2 (6 \cdot 31^2 - 0)$$

$$= 0 \cdot 0138$$

This is about 1·4% of the α-particles incident on the foil. The maximum thickness of gold foil which a 5 MeV α-particle can penetrate before losing all its kinetic energy in electron interactions is $4 \cdot 37 \times 10^{-4}$ cm, so that the thickness of the above scatterer is about one-quarter of the range of the α-particles. It is clear that only a very small fraction of the incident α-particles will be backscattered through large angles, and that many of these will have had their energy partially degraded due to partial absorption within the foil.

It is not possible to study the scattering of α-particles experimentally with sources whose activity is less than 1 mCi, which is well above the activity allowed for a single source by the Department of Education and Science regulations. The experiment is also complicated by the fact that it must be conducted in a vacuum to avoid the air absorption of the α-particles. The mechanism of α-particle scattering, however, can be studied with the aid of an analogy which is, in fact, capable of a more direct interpretation. The analogy allows both the scattering angle and the impact parameter to be measured directly and the value of b, which is related to the nuclear

Experiment 10 157

diameter, to be calculated. This is, of course, not the same as in an actual scattering experiment in which the correctness of the calculated value of b is inferred by showing that the distribution of α-particles with scattering angle agrees with that which is predicted theoretically. The analogy has the further advantage of simplicity in that it is essentially two-dimensional, whereas real scattering is complicated by the fact that it is a three-dimensional phenomenon.

The apparatus used to simulate α-particle scattering is shown in Plate 9. The α-particle is represented by a ball bearing which is given various kinetic energies by accelerating it down a ramp from selected heights. The nucleus is represented by a circular ramp which is shaped so that its height at any point is proportional to $1/r^2$, where r is the distance from that point to the centre of the ramp. In other words, it has the same cross-sectional shape as the Coulomb barrier surrounding the nucleus. The ball bearing will climb the circular ramp to a height which is determined by its kinetic energy and its impact parameter. For zero impact parameter, the height will be that at which the kinetic energy of the ball bearing has been completely transformed to potential energy, and at this point the diameter of the ramp will correspond to the apparent diameter of the nucleus for scattering of an α-particle (ball bearing) of that kinetic energy. It is possible to demonstrate the important concept that this apparent diameter is dependent upon the kinetic energy of the ball bearing or α-particle as is indicated for the α-particle in Fig. E.10.1.

Experimental

1. Place the accelerating ramp about 10 cm from the edge of the circular ramp and fix a stop on the ramp so that the ball bearing may always be released from the same position. This ensures that it will always gain the same kinetic energy. Plot the path of the ball bearing for various positions of the ramp so that different impact parameters are obtained. The path of the ball bearing may be plotted by chalking its surface so that it will record its track on a dark bench surface. Two or more sets of results may be obtained for ball bearings accelerated from different heights on the ramp. The kinetic energies will be directly proportional to the heights from which the ball bearings were accelerated.

2. After obtaining several tracks for a given ball-bearing kinetic energy, mark the position of the circumference of the circular ramp, remove the ramp and find the centre of the circle thus obtained. Then measure the impact parameter and scattering angle for each track, from the incident and scattered straight-line paths of the ball bearing by analogy with Fig. E.10.2. Record the height of centre from which the ball bearing was released on the accelerating ramp.

3. Using eqn. (E.10.11), calculate b for each scattering event and then average the values of b obtained for each kinetic energy used.

4. Show that b is approximately proportional to the reciprocal of the kinetic energy of the ball bearing. By analogy, the distance of closest approach of an α-particle to the nucleus is inversely proportional to its kinetic energy.

Note. Some results obtained with such an apparatus are given in Table 22. Clearly, this simple apparatus cannot yield highly con-

Table 22 Some results for α-particle scattering analogy

Measured value of p (cm)	Calculated value of b (cm)	Relative kinetic energy of ball bearing
7·0	3·28	1·0
6·0	3·88	1·0
4·4	3·32	1·0
2·9	2·76	1·0
2·2	3·28	1·0
	Av. 3·32	
9·1	2·88	1·49
6·2	1·84	1·49
2·8	3·40	1·49
2·4	2·20	1·49
1·7	2·52	1·49
1·0	3·20	1·49
0·8	2·32	1·49
0·7	1·24	1·49
	Av. 2·44	

Ratio b (k.e. 1·0) to b (k.e. 1·49) = 3·32/2·44 = 1·36

sistent results. The value of b obtained for a given kinetic energy is somewhat variable, the variation being greater when the higher kinetic energies are used. Nevertheless, the average value of b obtained certainly decreases as the kinetic energy of the ball bearing is increased, and the ratio of the two b values obtained is reasonably close to the inverse ratio of the kinetic energies.

The apparatus may, of course, be used in a purely qualitative way to demonstrate the dependence of the angle of scattering on the impact parameter.

Experiment 11
The absorption of gamma radiation

Gamma rays are emitted as a result of transitions between the energy states of a nucleus (see Chapter 1). They are electromagnetic radiations similar in nature to X-rays but are of much shorter wavelength. Like all electromagnetic radiations they travel with the velocity of light, and the energy, E, of the γ-ray photon is related to its frequency, ν, and its wavelength, λ, by the equation

$$E = h\nu = hc/\lambda$$

where h is Planck's constant and c is the velocity of light. Substituting numerical values for h and c in this equation,

$$E = \frac{1 \cdot 2414 \times 10^{-3}}{\lambda} \text{ mega-electron volts}$$

where λ is measured in nanometres (nm). The wavelength of a 1 MeV γ-ray is therefore 0·001 24 nm, and that of the 8 keV characteristic X-rays from a copper target is 0·154 nm. The familiar optical emission lines of sodium, with a wavelength near 170 nm, have a photon energy of about 7·4 eV.

In passing through matter, γ-photons may undergo absorption by three kinds on interaction, namely photoelectric absorption, Compton scattering and pair production. The mechanisms of these interactions have been discussed in Chapter 3. Generally, photoelectric absorption, and pair production do not lead to the production of scattered γ-rays, but in Compton scattering there is angular scattering of the γ-ray photons. If a narrow beam of γ-rays passes through an absorber the Compton scattered photons, since they are angularly scattered, do not enter a detector placed in the path of the collimated beam. The transmitted intensity, I, of γ-photons recorded by the detector then follows the law

$$I = I_0 \exp(-\mu x) \tag{E.11.1}$$

where μ is the linear attenuation coefficient (cm^{-1}), x is the linear thickness (cm) of the absorber, and I_0 is the intensity of γ-photons incident upon the absorber. The attenuation coefficient, μ, is the sum of partial coefficients for the three absorption processes involved; thus

$$\mu = \mu_{pe} + \mu_C + \mu_{pp}$$

Eqn. (E.11.1) may be rewritten as

$$I = I_0 \exp\left(-\frac{\mu}{\rho}\rho x\right)$$

or

$$\log_e I/I_0 = -\frac{\mu}{\rho}\rho x \tag{E.11.2}$$

where ρ is the density of the absorbing material. The quantity μ/ρ

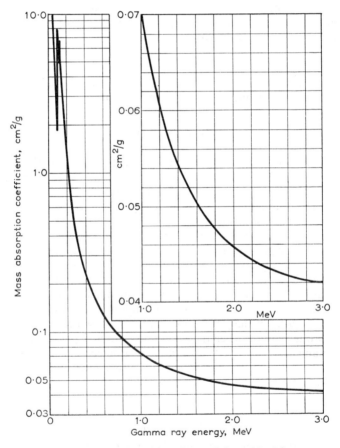

Fig E.11.1 Mass attenuation coefficient of lead for γ-ray energies up to 3·0 MeV

Experiment 11 161

(cm²/g) is called the *mass attenuation coefficient*, and ρx (g/cm²) is the thickness of the absorber expressed as mass per unit area.

The half-thickness of an absorber for a particular γ-ray energy is defined as the thickness required to reduce the transmitted intensity of the γ-rays to half of their incident intensity. Substitution in eqn. (E.11.2) yields the mass half-thickness:

$$\rho x_{1/2} = 0.693\rho/\mu \quad \text{grams/centimetre}^2$$

from which the linear half-thickness is

$$x_{1/2} = 0.693/\mu \quad \text{centimetre}$$

Experimental

1. Place a γ-ray source with an activity of several microcuries on the lowermost shelf of a Geiger–Müller counter castle. If β-particles are also emitted by the source, cover it with an aluminium absorber of sufficient thickness to absorb the maximum energy β-particles. (see Experiment 7)

2. With good statistical accuracy, determine the counting rate due to the source, first with no absorber present, and then successively with lead absorbers up to a few centimetres thick placed above it. Correct the counting rates for paralysis time and counter background.

3. Plot $\log_{10}$ (corrected counting rate) against lead absorber thickness expressed in g/cm². From the slope of the straight line so obtained calculate the mass attenuation coefficient, the linear attenuation coefficient and the linear half-thickness of lead for the γ-rays emitted by the source. The density of lead is 11·35 g/cm³.

4. Determine the energy of the γ-rays emitted by the source using the graph in Fig. E.11.1, which relates the mass attenuation coefficient to the photon energy.

Experiment 12

The inverse square law

We have seen in Experiment 11 the manner in which the intensity of collimated γ-rays decreases on passing through an absorber. In this experiment, the way in which the measured intensity of the γ-rays from an uncollimated source depends upon the distance from the source to the detector will be investigated. Gamma rays travel in straight lines with the velocity of light, and it will be found that their intensity decreases as $1/d^2$, where d is the distance from the point source to the detector. The intensity of γ radiation, I, at any point may be defined as the number of γ-photons passing through a

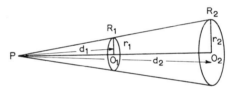

Fig E.12.1 The divergent cone of γ-rays from a point source

section of area 1 cm² perpendicular to the direction of travel at that point. The relationship, known as the *inverse square law*, may thus be stated as

$$I \propto \frac{1}{d^2}$$

and the law may be proved as follows. In Fig. E.12.1 a point source of γ-rays is situated at P. Consider the cone of γ-rays whose axis is PO_1O_2 and which at distance d_1 from the source passes through a circle of radius r_1. Since the γ-rays travel in straight lines, at distance d_2 from the source they will pass through a circle of radius r_2. Since the triangles PO_1R_1 and PO_2R_2 are similar,

$$r_1/d_1 = r_2/d_2$$

and therefore r is proportional to d. The intensity, I, of γ rays at distance d will be given by

$$I = \frac{A}{\pi r^2} \propto \frac{A}{\pi d^2} \propto \frac{1}{d^2}$$

Experiment 12

where A is the total number of photons in the cone of γ-rays and is constant for any particular source.

Experimental

It is easiest to carry out the experiment in the horizontal plane. The Geiger–Müller counter should be mounted so that the plane of its end window is vertical, and the source should move along a metre rule laid horizontally on the bench.

1. Place a γ-ray source of activity 5 to 50 μCi axially at a distance d from the counter window. Determine the counting rate at various values of d, say 10, 20, 40, 60 cm. The counting time in each case must be long enough to obtain a satisfactory statistical accuracy. The use of sources of lower activity will require longer counting times. (The Department of Education and Science approved γ-ray sources have an activity of 5 μCi.)

2. Correct the counting rate for counter background and plot the corrected counting rate against $1/d^2$. Alternatively, it may be shown that the plot of $1/\sqrt{n}$, where n is the counting rate, against d is linear. This overcomes the difficulty of deciding precisely where the counter zero should be taken.

Experiment 13

Observation of nuclear particle tracks in cloud chambers

A cloud chamber depends for its operation on the existence of a region of supersaturation in a gas which contains a condensable vapour. Ionizing radiations which pass through this region leave a trail of ions which act as centres for the condensation of the supersaturated vapour, thus producing a visible track. The *diffusion* and *expansion* types of cloud chamber differ only in the methods which are employed to obtain the supersaturation.

CONTINUOUS OR DIFFUSION-TYPE CLOUD CHAMBER
In the diffusion type of cloud chamber a vapour is evaporated at a warm surface and condensed on a cold surface, parallel to the warm surface and a few centimetres away from it. The gas just below the warm surface is saturated with the vapour, and as this vapour diffuses towards the cold surface the gas will become supersaturated. At or close to the cold surface the vapour will condense. Once this vapour concentration gradient has been established, the region of supersaturation will be stable and the chamber will be continuously sensitive to the passage of radiations through this region. To prevent an accumulation of ions which would cause condensation and thus destroy the supersaturation region, an electric field of about 20–50 V/cm is applied across the chamber. An adequate field may often be obtained by a static charge on insulating parts of such a chamber.

Fig. E.13.1(a) is a line drawing of a cloud chamber of this type. The upper Perspex disc forms the warm surface, and the felt pad attached to it is saturated with a 1:1 solution of ethyl alcohol in water*. The lower metal plate forms the cold surface and is cooled by placing it in contact with solid carbon dioxide. The tracks from an α-emitting source can readily be observed. Somewhat more difficult is the observation of β-particle tracks from β-particles emitted from a few milligrams of a uranyl salt wrapped in aluminium foil. This generally requires a larger chamber and good lighting. Note that the β-tracks can cross the chamber and are

* Iso-propyl alcohol may be substituted for this solution with advantage.

Experiment 13

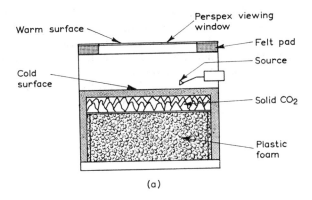

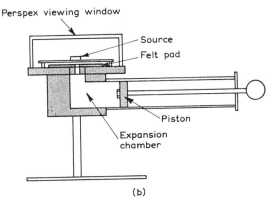

Fig. E.13.1 Demonstration cloud chambers

(a) Diffusion type   (b) Expansion type

occasionally deflected from straight paths due to collisions. The α-tracks are short, straight and much denser.

WILSON CLOUD CHAMBER (expansion cloud chamber)

In the expansion type of cloud chamber a gas is initially saturated with a condensable vapour. The gas is then rapidly expanded adiabatically, resulting in supersaturation. As heat flows into the chamber this supersaturation persists, from a few milliseconds to 2 or 3 seconds depending on the particular design. An electric field to prevent the build-up of ionization is again necessary.

Fig. E.13.1(b) shows the essential details of such a chamber. To observe tracks, remove the Perspex cover and saturate the felt pad with a 1:1 solution of ethyl alcohol in water*. Place an α-particle source in the chamber, replace the cover and connect a 120 V battery across the electrodes. Pull the piston sharply to its fullest extent and hold it in that position. The α-particle tracks should be visible, and the fact that α-particles travel in straight lines and all have the same range should be observed. Plate 10 shows the tracks due to α-particles from a radium-226 source in an expansion cloud chamber. The longer range α-particles are due to polonium-214, a decay product of radium.

Experiment 14
Gamma-ray scintillation spectrometry

The principles of scintillation counters and the instrumentation used in conjunction with them have been discussed in Chapter 3. The most frequently used phosphor for γ-ray counting is a single crystal of sodium iodide, activated with thallium impurity. Gamma-ray photons undergo various absorption processes in matter and in the crystal in particular, which result in the production of ionization in the absorbing medium. These absorption processes are the photoelectric effect, Compton scattering and pair production.

The photoelectric effect is an event in which the γ-photon transfers all its energy to an inner orbital electron in an atom of the absorber. The photon disappears and an electron is ejected from the atom. This scattered electron then undergoes absorption in the absorber.

Compton scattering may occur when a γ-ray photon interacts with an outer orbital electron of an atom in the absorber. The photon is scattered with a longer wavelength and part of the energy of the incident photon is transferred to the electron as kinetic energy. The scattered electron is readily absorbed if released in a dense absorber such as sodium iodide, but the scattered photon may or may not undergo further interactions in an absorber of limited dimensions.

Pair production is an event which occurs in the high electric field close to the nucleus of an atom. The γ-ray photon disappears and a positive and negative electron pair is created. This can only occur for γ-ray photons of minimum energy 1·02 MeV, the energy equivalent of the rest mass of the electron pair created. Any energy in excess of 1·02 MeV is partitioned as kinetic energy between the two particles. The electron produced in pair production soon loses its kinetic energy in the absorber and then becomes a part of the general electron population in the absorber. The positron, however, after losing its kinetic energy undergoes annihilation with an electron in the absorber. The two electrons involved disappear and two γ-photons, each of energy 0·51 MeV, are emitted in opposite directions. These annihilation photons are unlikely to be reabsorbed in the crystal, except for very large crystals, and give rise to the escape peaks mentioned below.

The energetic electrons resulting from these primary absorption processes dissipate their energy in the sodium iodide crystal and cause excitations throughout the crystal structure. Light photons are subsequently emitted as these excitations return to the ground state. The number of light photons produced, and hence the pulse size from the scintillation counter, is directly proportional to the energy lost by the γ-ray photon. A γ-ray spectrum is a graph of the frequency of occurrence of pulses against pulse size, and that shown in Fig. E.14.1 is the spectrum for a caesium-137 source obtained

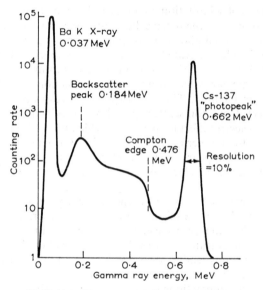

Fig E.14.1 The γ-ray spectrum of caesium-137

with a 2 in × 2 in NaI(Tl) crystal γ-ray spectrometer. It shows many of the common features of a γ-ray spectrum, which are defined below.

The *total energy peak* (commonly called the "photopeak") is due to the following phenomena.

1. The total energy of a γ-photon is absorbed in the scintillator by a primary photoelectric event, followed by absorption within the scintillator of any resulting scattered electrons and X-rays.

2. The total energy of a γ-photon is absorbed in the scintillator by a primary Compton scattering event, followed by absorption within the scintillator of the scattered electron and the scattered photon. The probability of such absorption increases with increasing crystal size.

3. In the case of a γ-photon of energy equal to or greater than 1·02 MeV, the total energy of the photon may be absorbed in the scintillator by a primary pair-production event, followed by absorption within the scintillator of the kinetic energy of both electrons and both the positron annihilation quanta. The probability of such absorption increases with increasing crystal size.

A *summation peak* may occur for radioisotopes which emit two (or more) coincident γ-rays. This peak corresponds to the sum of the γ-ray energies and arises because of the simultaneous absorption of both quanta in the scintillator. The probability of observing it increases with increasing crystal size.

With γ-rays of energy equal to or greater than 1·02 MeV, *escape peaks* may be observed at 0·51 MeV and 1·02 MeV less than the γ-photon energy. These occur when one or both of the annihilation quanta (following pair production and subsequent positron annihilation) are not absorbed in the scintillator. The probability of observing such escape peaks increases as the γ-ray energy increases (because of the increased probability of pair production), and decreases, especially for the double escape peak, as the crystal size is increased.

The *backscatter peak* occurs when γ-rays are back-scattered into the scintillator after undergoing a Compton scattering event in the materials surrounding the scintillator. The energy of the backscatter peak is approximately equal to that of a 180° Compton scattered photon, and this approaches a maximum of 0·25 MeV as the energy of the incident γ-ray increases.

The *Compton edge* arises because of the maximum in the number/energy distribution of Compton scattered electrons which occurs at an energy corresponding to 180° Compton scattered photons. Assuming that the scattered photon from a Compton scattering event within the scintillator escapes from the crystal without absorption, then a pulse will result which corresponds in energy to difference in energy between the γ-photon and the Compton scattered photon.

The energies of the backscatter peaks and Compton edges for various incident γ-ray energies are given in Appendix 10.

The *resolution* is defined as the width of the total energy peak at half its maximum height, expressed as a percentage of the peak energy. For a given crystal-photomultiplier assembly the resolution is approximately proportional to $1/\sqrt{E}$, where E is the energy of the γ-ray.

The *photofraction* is the fraction of the total number of pulses occurring in the spectrometer which lie in the total energy peak. The photofraction increases with increasing crystal size and decreasing γ-ray energy, coming close to unity for energies less than 0·2 MeV and medium-size crystals.

The *intrinsic efficiency* is the ratio of the number of pulses produced in the spectrometer to the number of γ-ray photons which enter the scintillator. It increases with increasing crystal size and decreasing γ-ray energy, though less markedly than does the photofraction. Its value is close to unity for γ-ray energies less than 0·2 MeV and medium-size crystals.

Some γ-ray sources which may be used for spectrometer calibration are listed in Table 23.

Table 23 Gamma-ray sources for spectrometer calibration

Source	Half-life	Gamma-ray energy, MeV	Abundance, %
Am-241	458 y	0·03	3
		0·06	36
Ba-133	10 y	0·081	36
		0·276	7
		0·302	14
		0·356	69
		0·383	7
Cs-137	30 y	0·03 (Ba X-ray)	8
		0·662	82
Co-57	270 d	0·014	8
		0·122	89
		0·136	9
Co-60	5·25 y	1·17	100
		1·33	100
Mn-54	314 d	0·84	100
Hg-203	47 d	0·07	13
		0·28	81
Na-22	2·6 y	0·51	181
		1·28	100
Y-88	106·5 d	0·9	92
		1·84	99·5
Th-228 (+ daughters)	1·91 y	0·084	2
		0·24	40
		1·62	10

All these sources are available from the Radiochemical Centre Ltd, Amersham.

Experimental

1. Select several γ-ray reference sources from Table 23 which have γ-ray energies that cover the energy range of interest.
2. Place the source with the highest γ-ray energy near the detector and adjust the amplifier gain so that no pulses of amplitude greater than the maximum which the pulse analyser can accept are obtained. For this purpose the analyser should be

Experiment 14 171

used in the discriminator mode, with the discriminator level set to its maximum. The amplifier gain is then adjusted so that pulses just fail to reach the scaler. The overall gain of the spectrometer may also be varied by making slight adjustments to the photomultiplier voltage applied, but the recommended operating voltage should not be greatly departed from.

3. Set the channel width to about 2% of the maximum discriminator level and set the pulse analyser to the analyser mode. Determine the counting rate due to the source as a function of the discriminator level, reducing the discriminator level by steps about equal to the channel width. Plot counting rate against discriminator level to obtain a graphical representation of the γ-ray spectrum of the source.
4. Without altering the spectrometer settings, similarly obtain spectra for the other sources.
5. Plot the discriminator levels at which the γ-photopeaks occur against the corresponding γ-ray energies. This calibration may then be used for the determination of the γ-ray energies, and hence the identification, of unknown sources.
6. Measure the photofraction on each of the spectra obtained.
7. Place a calibrated point γ-ray source close to the surface of the spectrometer crystal so that the geometrical efficiency may be assumed to be 50%. Obtain a spectrum as before and from it determine the photofraction and the intrinsic efficiency.
8. Measure the photopeak resolution for each of the spectra obtained and plot the resolution against $1/\sqrt{E}$ for each peak.
9. Determine the photopeak heights for several sources of known relative activities of the same radionuclide, and plot the photopeak height against the relative activity for each source.
10. Using spectrometers with different crystal sizes, investigate the effect of crystal size on photofraction and intrinsic efficiency for γ-rays of energies of about 0·1 MeV and greater than 1·0 MeV.

Experiment 15
An intercomparison of detector efficiencies

Detector efficiency may be defined as the percentage of the disintegrations occurring in a source that result in counts in the detector. For a given detector, this quantity will vary with the nature of the particles or photons being counted and with their energy. It is important to select a detector with a suitable efficiency for any given counting application. End-window Geiger–Müller counters have a high efficiency for the detection of α-particles and β-particles but a much lower efficiency for γ-ray photons. A Geiger–Müller counter with a glass window (such as a liquid sample counter) will not detect α-particles or low-energy β-particles at all. A scintillation counter with a NaI phosphor detects γ-rays with a much higher efficiency than a Geiger counter. It will also detect (with low efficiency) β-particles which have sufficient energy to penetrate the aluminium can of the NaI crystal. Low-energy β-emitters such as carbon-14 may be counted with high efficiency by a liquid scintillation counter (see Chapter 3) and with low efficiency by an end-window Geiger–Müller counter.

Experimental
The table below indicates the source activities which are suitable for counting by Geiger–Müller counters and scintillation counters.

Isotope	Geiger-Müller counter		Scintillation counter	
	End-window	Glass window for liquid samples	NaI phosphor	Liquid scintillator
Co-60 or I-131	0·1 μCi (A+)	10 cm³ of 0·01 μCi/cm³ (B)	0·1 μCi (A and C)	—
P-32	0·01 μCi (A)	10 cm³ of 0·01 μCi/cm³ (B)	0·01 μCi (A)	0·01 μCi (D)
C-14	0·01 μCi (A)	—	—	0·01 μCi (D)

Source forms A. Activity indicated on planchette with carrier added.
B. Activity indicated in solutions with carrier added.
C. As B but contained in stoppered sample bottle (polythene or glass).

D. Activity indicated in solution added to liquid scintillator solution (see Table 4).

+ These sources will emit β-particles as well as γ-rays, but the β-particles can be removed by covering the source with a 200 mg/cm² aluminium absorber.

1. Prepare sources for end-window Geiger–Müller counting by evaporating to dryness a solution containing the appropriate activity on a planchette. Add about 0·1 mg of carrier to the planchette before the evaporation.

Prepare sources for liquid-sample Geiger–Müller counting by preparing solutions with the indicated activity. Add about 0·1 mg/cm³ of carrier to the solution.

Radioisotope solutions supplied by the Radiochemical Centre Ltd., Amersham, are generally within $\pm 10\%$ of the nominal activity.

2. Determine the counting rates of the prepared sources as indicated in the table. Correct the counting rates as necessary for background and paralysis time.

3. Calculate the efficiencies of each counter from the calculated disintegration rate for each source and the measured counting rate.

The relative efficiencies of the various detectors for a given radioisotope may be determined to within the limits set by statistical counting errors. The absolute values of the efficiencies, however, will also be subject to the $\pm 10\%$ uncertainty in the activity of radioisotope solutions and to the errors of volumetric measurements. For more accurate work, standardized radioisotope solutions are available from the Radiochemical Centre.

Experiment 16
Self-absorption in beta particle sources

Some of the β-particles which are emitted by a radioactive source of finite thickness will be absorbed within the source. If a detector is placed above the source, then all the β-particles which leave the upper layers of the source and enter the detector will be counted. Those β-particles, however, which originate deeper in the source will suffer absorption in the upper layers of the source and may not reach the detector. This is the phenomenon known as *self-absorption*, and it is important to correct measured counting rates for its effect, particularly with those isotopes, such as carbon-14 or sulphur-35, which emit β-particles of low energy.

The observed counting rate of a source in which there is self-absorption will be dependent upon the source thickness. If the source is made sufficiently thick, β-particles emitted by the lowermost layers of the source will be completely absorbed by the source material lying above them. Fig. E.16.1 shows the variation of the counting rate of a source in which there is self-absorption with the thickness of the source. Curve A was obtained with sources prepared from carbon-14-labelled barium carbonate which had a specific activity equal to twice that of the material used for curve B. It is clear that the counting rate for each source material increases to a constant value as the source thickness is increased. Sources which are of sufficient thickness to reach this constant value are referred to as *infinitely thick* sources. The addition of further labelled material to such a source will increase its thickness, but the source counting rate will remain constant as only the upper layers of the source are contributing to the count. The minimum source thickness which may be regarded as infinitely thick is independent of the specific activity of the source material but varies for source materials of different chemical compositions. It may also be noted from inspection of Fig. E.16.1 that, for sources of the same chemical composition, the counting rate at infinite thickness is directly proportional to the specific activity of the source material.

The self-absorption of low-energy β-particles complicates the determination of the relative activities of sources which are prepared in many tracer experiments. In the determination of solubility, for

Experiment 16

example, it may be necessary to compare the activity of a thick precipitate with that of the very small amount of the same material which is deposited from a saturated solution (see Experiment 17.2). Many biological experiments involve the measurement of the relative activities of carbon-14-labelled barium carbonate precipitates (see, for example, Experiment 28.1). The sources which are

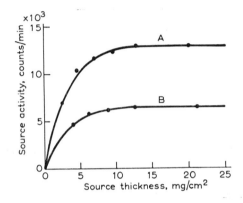

Fig E.16.1 Source activity plotted against source thickness for sources with self-absorption

The specific activity of source A was twice that of source B

prepared in such experiments will generally be of different thicknesses. Before their activities can be compared, therefore, it is necessary to correct each counting rate for the effect of self-absorption. The methods by which these corrections may be made are described empirically below. The basis of these methods is then established theoretically for those readers who are interested in a more formal treatment.

APPARENT SPECIFIC ACTIVITY AND TRUE SPECIFIC ACTIVITY

The *specific activity* of a labelled substance is the activity of unit mass of the material. For a source in which there is self-absorption the *apparent specific activity* is the source counting rate divided by its mass. The apparent specific activity will decrease as the source thickness is increased. Fig. E.16.2 shows the variation of the apparent specific activity with source thickness for the results used to obtain curve A in Fig. E.16.1. The apparent specific activity curve may be extrapolated to zero thickness and the intercept on the ordinate

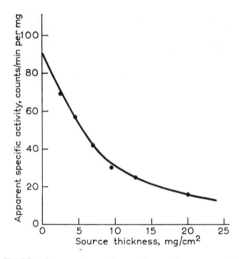

Fig E.16.2 Apparent specific activity against source thickness

will be the *true specific activity* of the source material since there will be no self-absorption in a source of negligible thickness.

PREPARATION OF A SELF-ABSORPTION CORRECTION CURVE

The true specific activity of the source material having been determined, the *true counting rate* for each prepared source thickness may be calculated. The true counting rate of a source is the counting rate it would have if there were no self-absorption, and is equal to the true specific activity multiplied by the source mass. The observed counting rate for each source may now be expressed as a percentage of its true counting rate. This will be called the *counting percentage* of the source. The counting percentage for the sources of Figs. E.16.1 and E.16.2 are plotted against source thickness in Fig. E.16.3. Once such a graph has been prepared it may be used to determine the counting percentage for any source of known thickness which has the same chemical composition as the sources used to prepare the graph. The true counting rate of the source may then be calculated. It is possible in this way to compare the true counting rates or specific activities of sources of different thicknesses. For accurate work, the sources must be counted with the same counter and under the same counter geometry conditions as those used for the preparation of the correction curve.

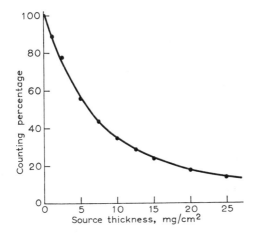

Fig E.16.3 Curve for making corrections for the effect of self-absorption in carbon-14-labelled barium carbonate sources

COMPARISON OF THE ACTIVITIES OF SOURCES OF INFINITE THICKNESS

For sources which are of less than infinite thickness it is necessary to make self-absorption corrections using a correction curve as described above. However, where sufficient source material is available for the preparation of infinitely thick sources, this is not necessary. As mentioned earlier, for an infinitely thick source, the counting rate is directly proportional to the specific activity of the source material. To compare the activities of infinitely thick sources it is only necessary to compare their counting rates as determined under identical counting conditions.

Experimental

Labelled material   Solution of carrier-free carbon-14-labelled sodium carbonate (Radiochemical Centre catalogue number CFA 2)

Recommended activity 1–5 μCi

An experimental study of self-absorption using a carbon-14-labelled barium carbonate precipitate is described below. The measurements are used to prepare a self-absorption correction curve which may be kept for permanent reference provided that the same counting arrangement is used.

1. To about 2·0 cm³ M sodium carbonate solution in a centrifuge cone add 1 to 5 μCi of carrier-free carbon-14-labelled sodium

carbonate solution and then quantitatively precipitate barium carbonate by dropwise addition of saturated barium chloride solution. Centrifuge the precipitate, test for complete precipitation and wash it several times with distilled water.

2. Prepare a series of barium carbonate sources of different thicknesses from a few milligrams on a planchette to "infinitely thick" sources.

For the preparation of a source, weigh a flat aluminium planchette (placed in a small watch-glass to facilitate handling and to prevent possible contamination of the balance) and transfer some precipitate, as a slurry with water or water and alcohol, to the planchette using a transfer pipette. Obtain as even a distribution of the source material on the planchette as possible. Dry the precipitate under an infra-red lamp, cool, and weigh the planchette plus watch-glass again.

3. Determine the counting rate of each source under the same geometry conditions and with good statistical accuracy. Correct each counting rate for counter paralysis time and for background.

Plot both the total activity and the apparent specific activity (i.e. counts/min per milligram of precipitate) of each source against the source thickness expressed in mg/cm^2 (i.e. milligrams of precipitate divided by the planchette area). Unit source thickness is 1 mg/cm^2.

Extrapolate the graph of apparent specific activity so obtained to zero thickness and thus determine the true specific activity of the barium carbonate precipitate. This is equal to the counting rate per milligram of precipitate at zero thickness.

4. Use the true specific activity so determined to calculate the true counting rate for each of the sources prepared. Then express the observed counting rate for each source as a percentage of its true counting rate, i.e. as a counting percentage. Plot these counting percentages against source thickness to obtain a graph similar to that of Fig. E.16.3.

This graph may be used to determine the counting percentage of any source of known thickness which is subsequently counted with the same detector and the same geometry. The true counting rate of such sources may then be calculated.

A THEORETICAL TREATMENT OF SELF-ABSORPTION

An expression may be derived for the observed counting rate of a source in which self-absorption occurs, on the assumption that the β-particles are absorbed by the source material in an exponential manner and that absorption is similar to that in external absorbers. Neither of these assumptions can, in fact, be completely justified, but a sufficiently accurate expression for the self-absorption effect can, nevertheless, be obtained. The mass absorption coefficient, μ, for β-particles is independent of the nature of the absorber for absorbers of low atomic number, provided that the absorber thickness is expressed in units of mass per unit area. The mass absorption coefficient of sodium carbonate, for example, will not be greatly different from that of aluminium. Many precipitates with which we are concerned, however, contain elements of high atomic number; for example, barium carbonate. As the mass absorption coefficient increases with atomic number, barium carbonate will have a high absorption coefficient. It is a reasonable assumption, however, that this absorption coefficient will be the same whether the β-particles are emitted within the precipitate or are transmitted through it from an adjacent carbon-14 source.

Consider a thin radioactive source of thickness x in which it can be assumed that there is no self-absorption. The observed counting rate of this source will therefore be the true counting rate, $n_{x\,\text{true}}$, of a source of thickness x. If the thickness of the source is now increased by an amount dx, by the addition of source material of the same specific activity, then the total activity of the source will be increased by an amount dn, where

$$dn = \frac{n_{x\,\text{true}}}{x} dx$$

As the thickness of the source is increased, however, self-absorption becomes important. The β-particles from the lowermost layer of thickness dx suffer absorption in the layer of thickness x above. The observed counting rate for the additional layer, dx, will therefore be

$$dn_{\text{observed}} = \frac{n_{x\,\text{true}}}{x} dx\, e^{-\mu x}$$

on the assumption that the absorption follows an exponential law. The observed counting rate for a source of finite thickness x is obtained on integration:

$$n_{x\,\text{observed}} = \int dn_{\text{observed}} = \int_0^x n_{x\,\text{true}} \frac{e^{-\mu x}}{x} dx$$

$$= \frac{n_{x\,\text{true}}}{\mu x}(1 - e^{-\mu x})$$

or

$$\frac{n_{x\,\text{observed}}}{n_{x\,\text{true}}} = \frac{1 - e^{-\mu x}}{\mu x} \qquad \text{(E.16.1)}$$

As the thickness x increases $(1 - e^{-\mu x})$ approaches unity, and for a source of "infinite thickness", n_{xi}, eqn. (E.16.1) reduces to

$$n_{xi\,\text{observed}} = \frac{n_{xi\,\text{true}}}{\mu x_i} = \frac{1}{\mu} \times \text{true specific activity} \qquad \text{(E.16.2)}$$

Since $1/\mu$ is constant, the counting rates of infinitely thick sources must be directly proportional to the specific activities of the materials of which they are composed, as was mentioned in the discussion of Fig. E.16.1.

Eqn. (E.16.1) shows that $n_{x\,\text{observed}}/n_{x\,\text{true}}$, or the counting percentage, is proportional to $(1 - e^{-\mu x})/\mu x$. This involves only the constant μ and the thickness of the source, x. Thus the counting percentage is independent of the specific activity of the source material. It is therefore valid to apply the correction curve of Fig. E.16.3 to all sources of barium carbonate, irrespective of their specific activities.

Experimental

The treatment of the results used to prepare Fig. E.16.3 may be extended to show that the experimental curve agrees closely with eqn. (E.16.1) and to determine the mass absorption coefficient, μ, of the precipitate for self-absorption of β-particles. The curve of Fig. E.16.3 should be of the same shape as a plot of $(1 - e^{-x})/x$ against x. Values of this expression for various values of x and of the theoretical counting percentage are given in Table 24.

Table 24 Theoretical values of counting efficiency for sources with self-absorption

$X = \mu x$, where μ is the mass absorption coefficient and x is the source thickness

X	$(1 - e^{-X})/X$	Counting efficiency, %
0·5	0·7870	78·7
1·0	0·6321	63·2
2·0	0·4323	43·2
3·0	0·3167	31·7
4·0	0·2457	24·6
5·0	0·1986	19·9
6·0	0·1662	16·6

1. Mark the theoretical counting percentage for $X = 5$ on the experimental curve. Divide the abscissa scale into five equal parts between zero and the abscissa value corresponding to this point. Use this new abscissa scale to plot the other theoretical counting percentage values against X. The theoretical and experimental curves so obtained should be identical.

Experiment 16

2. Read the source thickness corresponding to a counting percentage of 19·9% and express this thickness in terms of mg/cm^2. This value of x must be related to X by the equation $X = \mu x$ or $5 = \mu x$ for 19·9% counting efficiency, from which μ can be calculated.

For the curve shown in Fig. E.16.3,

x at 19·9% counting percentage = 21 mg/cm^2

and

$X = 5 = 0·021\mu$ or $\mu = 5/0·021 = 238 \ cm^2/g$

Experiment 17

Determination of the solubility of a slightly soluble salt

Most salts which are regarded as being insoluble do in fact dissolve in water to some extent. The solubilities observed vary from a few micrograms per 1000 cm^3 to a few hundred milligrams per 1000 cm^3. Calculation shows that a saturated solution of an "insoluble" salt will therefore contain a minimum of about 10^{12} molecules in each cubic centimetre of solution. In the radiochemical method for determining solubility a saturated solution is prepared from a precipitate of the salt which has been labelled with a radioactive tracer. Such numbers of radioactive atoms can quite readily be detected. For example, even if the tracer used has a half-life as long as one year, each cubic centimetre of the saturated solution will have an activity of a few thousand disintegrations per minute. If the specific activity of the labelled salt is known, then the solubility may be calculated from the measured activity of the saturated solution. Either the labelled salt may be prepared in such a way that its specific activity is known, or alternatively its specific activity may be determined after preparation. In Experiment 17.1, silver-110m-labelled silver chloride is prepared in the former manner and the solubility of silver chloride in water and sodium chloride solutions is determined. Experiment 17.2 describes the preparation of a carbon-14 labelled strontium carbonate precipitate and the subsequent determination of its specific activity, and the measurement of the solubility of strontium carbonate in water.

SOLUBILITY AND SOLUBILITY PRODUCTS

A saturated solution of a salt BA will contain the ions B$^+$ and A$^-$ which will be in equilibrium with the solid phase due to the dissociation

$$BA \rightleftharpoons B^+ + A^-$$

The concentration of ions in the solution is determined by the solubility product, S_{BA}, which is defined as

$$S_{BA} = [B^+][A^-]$$

The solubility, s, of each ionic species is equal to its concentration in the solution and is given by the equation

$$s_B = s_A = (S_{BA})^{1/2}$$

for a solution of the salt in water. If the salt is dissolved in a solution of a soluble salt which contains either of the ions B^+ or A^-, then the ionic solubilities are given by

$$s_{B^+} = S_{BA}/[A^-] \text{ in a solution of a salt containing } A^-$$

and

$$s_{A^-} = S_{BA}/[B^+] \text{ in a solution of a salt containing } B^+$$

The ionic solubility is therefore reduced as the concentration of the oppositely charged ion in the solution is increased. The concentration of silver chloride, for example, is less in dilute sodium chloride solutions than in water.

The solubility of silver chloride in sodium chloride solutions begins to increase again for chloride concentrations greater than 0·01 M. This increase is due to the formation of soluble complexes between the silver and chloride ions. At sodium chloride concentrations up to about 2·0 M the predominant complex ion formed is $[AgCl_2]^-$. This ion will be formed by the reversible reaction

$$Ag^+ + 2Cl^- \rightleftharpoons [AgCl_2]^-$$

The extent to which it will be formed is dependent upon the square of the chloride concentration, and the stability constant for the complex is defined as follows:

$$\text{Stability constant} = \frac{[AgCl_2]^-}{[Ag^+][Cl^-]^2}$$

If the chloride concentration is increased beyond 2·0 M, higher complexes such as $[AgCl_3]^{2-}$ also contribute to the solubility of silver. The silver solubility therefore increases to an even greater extent than would be expected from the dependence upon the square of the chloride concentration below 2·0 M chloride concentrations.

EXPERIMENT 17.1 DETERMINATION OF THE SOLUBILITY OF SILVER CHLORIDE IN WATER AND IN SODIUM CHLORIDE SOLUTIONS

The solubility of silver chloride can easily be determined using silver-110m-labelled precipitates of silver chloride. Silver-110m has a half-life of 253 days and decays principally by β-emission followed by the emission of γ-radiation. The β-particle energies are 0·085 and

0·530 MeV. The most important γ-ray energies are 0·66 and 0·89 MeV, but there are many others from 0·45 to 1·56 MeV. The β-particles emitted will not penetrate the glass window of a liquid-sample Geiger–Müller counter, but it is possible to measure silver-110m activities with such a counter because of the abundant emission of γ-rays. The solubility of silver chloride in sodium chloride solutions less than 0·5 M in concentration is very low, and it is therefore advisable to use a precipitate of higher specific activity than is necessary for determinations of solubility in greater than 0·5 M sodium chloride solutions.

Experimental

Labelled material   Silver-110m-labelled silver nitrate solution (Radiochemical Centre catalogue number SES 1)
Recommended activity 20 μCi

1. Prepare a 0·01 M silver nitrate solution by dissolving 0·17 g of silver nitrate in 0·1 M nitric acid in a 100 cm³ volumetric flask.

2(*a*). *For the determination of the solubility in water and in sodium chloride solutions of concentration less than 0·5 M.*

Pipette 5·0 cm³ of 0·01 M silver nitrate solution into a centrifuge cone and add 20 μCi carrier-free silver-110m solution. Add 0·1 M hydrochloric acid dropwise until precipitation of silver chloride is complete.

2(*b*). *For the determination of the solubility in sodium chloride solutions of concentration greater than 0·5 M.*

Pipette 20 cm³ of 0·01 M silver nitrate into a centrifuge cone and add 20 μCi carrier-free silver-110m solution. Add 0·1 M hydrochloric acid dropwise until precipitation of silver chloride is complete.

3. Spin down and wash the precipitate several times with distilled water. Finally remove as much as possible of the last water wash with a transfer pipette.

4. Add 2 μCi of carrier-free silver-110m to a few cm³ of 0·01 M silver nitrate in a 100 cm³ volumetric flask, and make up to volume with distilled water. Determine the activity of 10 cm³ of this solution in a liquid Geiger–Müller counter.

5. *For* (*a*). To 20 cm³ of distilled water and 0·005 M, 0·05 M and 0·5 M sodium chloride solutions in small stoppered bottles, add sufficient of the labelled precipitate to saturate each. Transfer the precipitate using a microspatula.

For (*b*). To 20 cm³ of 0·5 M, 0·75 M, 1·0 M, 1·5 M and 2·0 M sodium chloride solutions, add labelled silver chloride precipitate as described above.

Experiment 17.2 185

6. Allow the solutions to stand for about 15 min with occasional shaking. Then filter or centrifuge the solutions to remove the excess silver chloride precipitate. Pipette 10 cm^3 of each solution into the liquid Geiger–Müller counter and determine the activity of each.

7. From 4, calculate the total activity used to prepare the precipitate and hence the specific activity of the precipitate. Then calculate the weights of silver chloride present in 10 cm^3 of each saturated solution from their activities measured under 6.

8. Calculate the silver chloride solubilities in moles/1000 cm^3. Plot the molar solubility against the square of the sodium chloride concentration.

Notes

1. The 20 μCi of silver-110m used in 2 and the 2 μCi used under 4 must be in exactly the correct ratio. This can be achieved by adding approximately 22 μCi of carrier-free silver-110m to about 12 cm^3 of distilled water in a small beaker. The 20 μCi may then be dispensed by accurately pipetting 10 cm^3 of this solution, and the 2 μCi by pipetting exactly 1·0 cm^3.

2. The counter background should be determined before each activity measurement. The counter should be filled with distilled water for the background measurement.

3. To minimize any difficulties that may arise due to contamination of the counter, the activities of the saturated solutions should be determined in the order of increasing solubility, and the activity of the silver-110m solution under 4 should be measured last.

4. The maximum activity of silver-110m which may be used in a single experiment under the Department of Education and Science regulations is 24 μCi. The solubility determinations for the two sodium chloride concentration ranges should not therefore be carried out as a single experiment unless the suggested activities are reduced

EXPERIMENT 17.2 DETERMINATION OF THE SOLUBILITY OF STRONTIUM CARBONATE

Experimental

Labelled material   Carbon-14-labelled sodium carbonate solution (Radiochemical Centre catalogue number CFA 2)
Recommended activity 10 μCi

(A) *Preparation of carbon-14-labelled strontium carbonate*

1. Add about 10 μCi of carbon-14-labelled sodium carbonate solution to 5 cm^3 of 0·1 M sodium carbonate solution in a 50 cm^3 centrifuge tube.

2. Quantitatively precipitate strontium carbonate by the addition of 0·1 M strontium chloride solution. Centrifuge the precipitate and test for complete precipitation by adding a little more strontium chloride solution.

3. Wash the precipitate several times with distilled water or until the washings are free from chloride.

(B) *Determination of the solubility of strontium carbonate*

1. Transfer some of the strontium carbonate precipitate to a small stoppered flask containing about 10 cm^3 of distilled water. Stopper and shake the flask for about 15 min to saturate the solution.

2. Filter the saturated solution into a dry flask using a dry filter paper and filter funnel.

3. To determine the carbon-14 activity present in the saturated solution prepare several sources by the following method. To 0·05 cm^3 of 0·1 M sodium carbonate solution placed in an aluminium planchette, add exactly 1·0 cm^3 of the saturated strontium carbonate solution. Evaporate this to dryness under an infra-red lamp whilst maintaining a stream of carbon-dioxide-free air over the surface of the source. An air stream from an aquarium aerator may be used, removing the carbon dioxide by passing it through 2 M sodium hydroxide solution.

Note. Dilute carbon-14-labelled carbonate solutions readily lose their activity by exchange with atmospheric carbon dioxide. The use of a carbon-dioxide-free air stream during drying and the addition of sodium carbonate carrier prevent such exchange. The amount of carrier added must be kept small so that the prepared source may be considered as of zero thickness.

4. Determine the counting rates due to the sources prepared from the saturated strontium carbonate solution. Use an end-window Geiger–Müller counter and obtain counts of good statistical accuracy. The counter background must also be determined accurately and subtracted from the source counting rates.

5. Use the remainder of the strontium carbonate precipitate to prepare sources of different thicknesses (up to 20 mg of precipitate on the planchette) and determine their counting rates under the same counting conditions that were used for the saturated solution. From these results determine the true specific activity of the strontium carbonate precipitate by the method described in Experiment 16. This procedure is necessary since carbon-14 emits a low-energy β-particle which suffers self-absorption in the precipitate.

6. Divide the average counting rate of the sources prepared from the saturated solution by the true specific activity of the precipitate to obtain the mass of precipitate required to saturate 1·0 cm^3 of

Experiment 17.2

solution. Calculate the solubility of strontium carbonate in g/1000 cm^3.

Further work. The solubility of barium carbonate may be determined by the same method, but as its solubility is only about one-tenth of that of strontium carbonate, about ten times as much carbon-14 activity must be used.

Experiment 18

Separation of radon-220 from the thorium series and determination of its half-life using an ionization chamber

The first part of the thorium series is as follows:

$$^{232}_{90}\text{Th} \xrightarrow{\alpha} {}^{228}_{88}\text{Ra} \xrightarrow{\beta} {}^{228}_{89}\text{Ac} \xrightarrow{\beta} {}^{228}_{90}\text{Th} \xrightarrow{\alpha} {}^{224}_{88}\text{Ra} \xrightarrow{\alpha} {}^{220}_{86}\text{Rn} \xrightarrow{\alpha} {}^{216}_{84}\text{Po} \longrightarrow$$

Half-life

1·39 × 10^{10} y   6·7 y   6·13 h   1·9 y   3·64 d   54·4 s   0·158 s

Chemical separation of thorium from mineral sources removes thorium-232 and thorium-228 in their equilibrium ratio. Equilibrium from thorium-228 onwards is established after about 20 days, but the total activity of this part of the series then decays with the 1·9 y half-life of thorium-228. To establish equilibrium from thorium-232 onwards requires about 35 years. Thorium salts which are sufficiently old to have largely re-established this equilibrium are sometimes described as "vintage" salts. Radon gas readily escapes from a finely divided thorium salt, especially thorium hydroxide. The efficiency of separation of radon can be increased by spreading the salt over a surface of lint or gauze, thus reducing the length of the diffusion path in the salt.

The half-life of radon can be determined by measuring its activity as a function of time as described in Experiment 1.

Experimental

(A) *Using a pulse electroscope*

1. Set up a pulse electroscope with an ionization chamber, and apply a potential sufficient to saturate the ionization chamber, via a safety resistor of about 10 MΩ, as shown in Fig. E.3.2.

2. Connect a polythene bottle containing thorium hydroxide to the inlet of the ionization chamber and squeeze the bottle several times to blow radon gas into the chamber.

Commercially available radon bottles have an outlet and an inlet tube with a non-return valve on the outlet. These tubes can be connected to inlet and outlet tubes on the ionization chamber to provide a closed system from which the radioactive gas cannot escape.

Experiment 18

3. Move the side electrode in towards the electroscope leaf until the frequency of movement of the leaf is as high as possible.

4. Pump some more radon into the chamber and record the time after every 5 to 10 discharges. When the discharge frequency has become low enough, record the time after each discharge.

5. Calculate the frequency of discharge events as a function of time. Plot $\log_{10}$ (discharge frequency) against time. The discharge frequency is proportional to the ionization current, which in turn is a measure of the activity of the radon. The half-life of radon may thus be obtained from the slope of the graph.

(B) *Using a quartz-fibre electroscope/ionization chamber* (page 137)

1. Pump some radon from a thorium hydroxide bottle into the ionization chamber and charge the electroscope.

2. Record the time required for the quartz fibre to move between two selected points on the scale. Note also the actual time at which this discharge begins.

Recharge the electroscope and repeat this operation several times.

3. Plot $1/t$ against T, where t is the time required to discharge between these fixed points and T is the actual mid-time of each discharge. Plot $\log_{10} 1/t$ against T and calculate the half-life of radon-220.

Experiment 19

Radiochemical separation of protoactinium-234m from the uranium series

The first few members of the uranium decay chain are shown in Table 25. Immediately after the chemical preparation of uranium

Table 25 Part of the uranium decay series

Radionuclide	Half-life	Particle energy, MeV
$^{238}_{92}\text{U}$	4.5×10^9 y	
$\downarrow \alpha$		4·18
(UX$_1$) $^{234}_{90}\text{Th}$	241·1 d	
$\downarrow \beta$		0·10
		0·19
(UX$_2$) $^{234}_{91}\text{Pa}$	1·18 min	
$\downarrow \beta$		2·31
(UII) $^{234}_{92}\text{U}$	2.5×10^5 y	
$\downarrow \alpha$		4·76
Ionium $^{230}_{90}\text{Th}$	8×10^4 y	
$\downarrow \alpha$		4·68
		4·61
$^{226}_{88}\text{Ra}$	1620 y	
$\downarrow \alpha$		4·78
		4·59

salts from mineral sources, uranium-238 and uranium-234 will be present in their secular equilibrium proportions. The growth of thorium-234 and protoactinium-234m into equilibrium with the uranium-238 parent then requires only a few months for completion, but equilibrium beyond uranium-234 will not be established in other than mineral sources. The isotopes thorium-234 and protoactinium-234m can be separated from uranium reagents as described below.

METHOD 1   SOLVENT-EXTRACTION/ION-EXCHANGE (UX$_2$ COW)

A cation exchange column may be used to separate uranium-238, thorium-234 and protoactinium-234m. The thorium and protoactinium are held more strongly on the resin than the uranium.

These decay products are, however, present in much smaller quantities than the parent uranium, and it is therefore necessary to use a large volume of cation exchange resin to effect the separation. It is subsequently difficult to separate the short-lived protoactinium-234m from the column before it decays. The size of ion exchange column required may be reduced by a preliminary separation of uranium by solvent extraction; the isolation of protoactinium-234m is then much easier.

Experimental

1. Place 15 g of crystals of uranyl nitrate, $UO_2 \cdot (NO_3)_2 \cdot 6H_2O$, in a separating funnel and add 30 cm³ of ether. Shake the funnel until the uranyl nitrate is completely dissolved and then for a further minute. The small aqueous layer formed by the water of crystallization contains most of the thorium.

2. Run the aqueous layer onto an ion exchange column (approximately 0·5 cm diameter × 5 cm long, Zeocarb 225 in H-form, 14–52 mesh).

3. Pass the following solutions through the column at a flow rate of about 1 drop every 2 s (about 2 cm³/min):

(a) 20 cm³ of distilled water
(b) 50 cm³ of 0·3 % sulphuric acid (14 cm³ of 2M sulphuric acid made up to 250 cm³)
(c) 20 cm³ of distilled water

The thorium-234 is strongly held by the resin and remains at the top of the column. Run 10 cm³ of M hydrochloric acid through the column and collect the eluate in a liquid counter.

4. Take 1 min counts of this eluted protoactinium solution as frequently as possible. Note the time at which each count starts and finishes.

5. Plot $\log_{10}$ (corrected counting rate) against the average time at which the count was taken. Determine the half-life of protoactinium-234 from the slope of the straight line obtained.

Note. On no account must the liquid level in the column be allowed to fall below the surface of the ion exchange resin. If it does, channels will form in the resin and the experiment must be started again. After separation of protoactinium-234, the column should be washed with 20 cm³ of distilled water. Protoactinium may be periodically "milked" from the column until the thorium-234 has decayed.

METHOD 2 SOLVENT EXTRACTION

Some organic solvents will selectively extract protoactinium from a solution of a uranyl salt in 7M hydrochloric acid. The higher alcohols and ketones have been found most effective, the iso-compounds giving a much higher extraction efficiency than the n-compounds.

Experimental

(A) *To observe the decay of separated protoactinium-234m*

1. Place about 100 cm³ of a 10% solution of uranyl nitrate in 7M hydrochloric acid (70 cm³ concentrated hydrochloric acid + 30 cm³ water) in a separating funnel.
2. Add about 15 cm³ of extractant (iso-butyl methyl ketone, or isoamyl acetate).
3. Shake the funnel for about 3 min and then, as quickly as possible, return the aqueous layer to its storage bottle and run the organic layer into a liquid counter.
4. Start a stopwatch and record half-minute counts as follows:

   minutes
   0·0 –0·5
   0·75–1·25
   1·5 –2·25 etc.

(B) *To observe the growth of protoactinium-234m in uranyl nitrate solution from which it has been separated*

1. Place about 15 cm³ of the uranyl nitrate solution in a separating funnel.
2. Add about 75 cm³ of the extractant.
3. Shake the funnel for about 3 min, start a stopwatch as the layers are allowed to separate and then run the aqueous layer into a liquid counter and start counting immediately, taking counts as above.
4. Later, return the organic layer to the solvent storage bottle for re-use.

In each case plot the count rate against average time at which each count was taken. In case (A), plot also the $\log_{10}$ (count rate) against the average time at which the count was taken and determine the half-life of protoactinium-234m.

METHOD 3 SEPARATION OF CARRIER-FREE THORIUM-234 AND PROTOACTINIUM-234m FROM THE URANIUM SERIES

The separation method outlined below demonstrates the behaviour of carrier-free thorium-234 as a radiocolloid and the use of a non-isotopic carrier for thorium. Thorium and protoactinium are the

Experiment 19

second and third members of the actinide series, and lanthanum, from the homologous lanthanide series, may be used as a carrier. Carrier-free thorium-234, in nearly neutral solutions, forms a radiocolloid which is probably thorium hydroxide formed by hydrolysis. This radiocolloid is strongly adsorbed on glassware on heating the solution, and hence separation from the lanthanum carrier can easily be effected.

A thorium/protoactinium separation may be effected by evaporating a concentrated nitric acid solution of thorium-234 and protoactinium-234m to dryness on a platinum planchette. The deposited protoactinium oxide is much more difficult to remove from the planchette than the thorium, which may then be removed by a rapid wash in concentrated nitric acid.

Experimental

1. Place 15 g of uranyl nitrate, $UO_2(NO_3)_2 \cdot 6H_2O$, in a separating funnel and add 30 cm^3 of ether. Shake the funnel until the uranyl nitrate is completely dissolved and then for a further minute. The small aqueous layer formed by the water of crystallization contains most of the thorium. Run the aqueous layer into a 50 cm^3 beaker, add 5 drops of concentrated nitric acid and evaporate until crystals start to separate. Cool, add 30 cm^3 of ether and stir. Decant the ether from the few drops of aqueous phase present.

2. Add 1 cm^3 of lanthanum carrier solution containing 5 mg of La^{+++}/cm^3 in 2% nitric acid. Evaporate the solution to dryness under an infra-red lamp.

3. Dissolve the residues in 25 cm^3 of water and add dilute ammonium hydroxide dropwise to raise the pH value to 9. Any uranium still present will be precipitated as ammonium diuranate, $(NH_4)_2U_2O_7$, which should be removed by filtration. The lanthanum carrier ensures that the thorium remains in solution. Heat the solution nearly to boiling for 1 hour, then pour the solution from the beaker and rinse it with a few cubic centimetres of distilled water.

4. Dissolve the adsorbed thorium from the beaker by washing it with 1 cm^3 of hot 2M nitric acid.

5. Prepare a thorium-234/protoactinium-234m source by evaporating 0·5 cm^3 of this solution to dryness on a platinum planchette. Determine the counting rate of this source using an end-window Geiger–Müller counter, covering the source with a 50 mg/cm^2 absorber so that only the protoactinium-234 β-particles are counted. The β-particle energies of thorium-234 and protoactinium-234m may be determined by the absorber method if desired.

6. To separate the thorium-234 and protoactinium-234m on the planchette, heat the planchette on a hotplate, and then add a few drops of concentrated nitric acid, sufficient to cover the area of

active deposit. Evaporate the nitric acid to dryness and heat on the hotplate for a further 30 s. Start a stop-clock at the instant when the nitric acid has been completely evaporated; this may be regarded as the instant of separation of the thorium-234 and protoactinium-234m. Immediately immerse the planchette in concentrated nitric acid for *1 s or less*, then quickly rinse in a beaker of distilled water, blot it dry with filter paper and finally dry quickly on a hotplate.

7. To determine the half-life of protoactinium-234m, record counts with an end-window Geiger–Müller counter as below. Cover the source with a 50 mg/cm² aluminium absorber so that the counting efficiency will be the same as for the thorium-234/protoactinium-234m source.

minutes
1·0 –1·5
1·75–2·25
2·5 –3·00
3·25–3·75 etc.

Continue taking counts at similar intervals until decay is complete.

Determine the half-life by plotting $\log_{10}$ (activity) against time and then calculate the protoactinium-234m activity present at the instant of separation. The efficiency of the separation may be obtained by comparing this count rate with that of the thorium-234/protoactinium-234m source before separation.

8. Calculate the theoretical masses of thorium and protoactinium which are being manipulated in these separations.

Experiment 20

Radiochemical separation of lead-212 and bismuth-212 from the thorium series

There are two methods by which the radiochemical separation of lead-212 and bismuth-212 from the thorium series can be effected. In the first method, lead and bismuth are separated electrolytically from a solution of thorium nitrate, whereas the second method makes use of the emanation (or escape) of radon gas from a solid in which it is produced by decay. Table 26 lists the decay sequence for

Table 26 The latter part of the thorium series

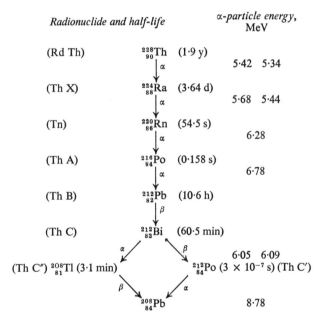

the latter part of the thorium series. Since all the half-lives involved are short, secular equilibrium from thorium-228 onwards will soon be established in a freshly prepared thorium hydroxide precipitate. If such a precipitate is dried and ground to a fine particle size, the radon-220 formed within it can readily diffuse into the surrounding air space. This radon rapidly decays to polonium-216 and then to lead-212. If the extra-nuclear electrons of polonium are not disturbed in this latter decay, we may expect the following charge distribution to ensue from the nuclear changes involved:

$$^{216}_{84}\text{Po} \rightarrow {}^{212}_{82}\text{Pb}^{--} + {}^{4}_{2}\text{He}^{++}$$

The emission of the α-particle, however, imparts a recoil momentum to the newly formed lead nucleus resulting in its ejection from some of its surrounding electrons. This process is known as *alpha recoil*, and it causes the lead atom to be produced as a positively charged ion instead of a negatively charged one. These positive lead ions are formed in the gas phase and may be collected on a negatively charged electrode.

Experimental

(A) *Separation of lead-212 and bismuth-212 from the thorium series by electrolysis*

1. Dissolve about 10 g of thorium nitrate in 25 cm³ 0·1M nitric acid. Partially immerse two 1 cm × 2 cm platinum-foil electrodes in this solution placed in a 50 cm³ beaker.

2. Connect a 6 V battery or other d.c. power supply across the electrodes through a rheostat with an ammeter in series. Adjust the current to 0·5 A and continue electrolysis for about 1 h. Lead-212 and bismuth-212 will be deposited on the cathode.

3. Place the electrolysed thorium nitrate solution in a storage bottle. Electrolysis may be repeated when the lead-212 has grown back into equilibrium in the solution, a minimum period of about 24 h being required.

(B) *Separation of lead-212 and bismuth-212 from the thorium series by an emanation method*

The apparatus used for the collection of the lead ions is shown in Plate 11. It consists essentially of a metal tray containing thorium hydroxide and a support for an aluminium-foil cathode.

1. Either use commercially available thorium hydroxide or prepare a finely divided precipitate by the following method. Dissolve 20 g of thorium nitrate in 200 cm³ of distilled water and precipitate thorium hydroxide by the slow addition of excess dilute ammonium hydroxide solution. Filter off the precipitate, wash it with distilled water and dry it overnight in an oven at 100°C. Grind the dried precipitate to a fine powder using a mortar and pestle.

Place some of the finely divided thorium hydroxide in the metal tray of the apparatus shown in Plate 11.

Experiment 20

2. Connect a 120 V battery across the electrodes, making the collector foil the negative electrode. The lead ions should be collected over a period of at least 24 h.

(C) *Determination of the β-energies of lead-212 and bismuth-212*

The β-particle energies of lead-212 and bismuth-212 may be determined using the platinum or aluminium foil sources prepared above. An absorption curve for the β-particles is plotted, and the range and energy are determined by the method described in Experiment 7.

1. Fix the aluminium or platinum cathodes on which the lead and bismuth isotopes have been collected onto a counting planchette with a little adhesive or cellulose tape.

2. Place the source on the second shelf of a Geiger–Müller counter castle and determine the counting rates when aluminium absorbers of the following thicknesses are placed on the shelf above the source: about 5, 10, 20, 40, 60, 80, 100 mg/cm^2 and then every 200 mg/cm^2 to 1500 mg/cm^2.

The length of the counting time for the thicker absorbers must be increased in order to obtain an acceptable statistical accuracy. With a 15 min counting time for the heaviest absorbers, the β-particle energy can generally be obtained to within about 10%.

3. Correct the absorber thicknesses for the thickness of the counter window and for the air space between the source and the counter window (add 1·3 mg/cm^2 for each centimetre of air distance). Correct the counting rates for counter dead time, but do not correct for the background counting rate, as this would increase the statistical errors present.

4. Plot $\log_{10}$ (counting rate) against absorber thickness and determine the range of the more energetic bismuth-212 β-particles by the extrapolation described in Experiment 7. Calculate the β-particle energy using the Glendenin equations.

5. The range and energy of the lead-212 β-particles may be obtained from the inflexion in the plot of $\log_{10}$ (counting rate) against absorber thickness which occurs at thicknesses of less than 100 mg/cm^2. Use the general shape of the curve above 100 mg/cm^2 to extrapolate it back to zero thickness. If this extrapolated curve is now subtracted from the early part of the experimental curve, a curve which extrapolates to the range of the lead-212 β-particles will be obtained. The energy of the β-particles may then be calculated from the Glendenin equations.

(D) *Separation of bismuth-212 and the determination of its half-life*

The removal of the lead-212 and bismuth-212 activities from the collecting foil and the subsequent separation of bismuth from the lead are described below.

1. Place the collecting foil in a test tube and add 2 cm³ of 0·3M nitric acid. Heat the solution to boiling for a few minutes, keeping the foil moving with a glass rod. Cool and transfer the solution to a centrifuge cone using a transfer pipette.

2. Add 3 drops of 1% bismuth nitrate carrier solution and 3 drops of 15% lead nitrate carrier solution. Then add 6 drops of M sulphuric acid to precipitate the lead quantitatively as lead sulphate. Centrifuge for a few minutes, and taking care not to disturb the precipitate, transfer the bismuth solution to a nickel-plated planchette.

3. Carefully warm the planchette (placed in a watch-glass) on a hotplate until metallic bismuth is deposited as a black film. This occurs due to electrochemical exchange between the nickel of the planchette and bismuth in the solution. As soon as the bismuth has been deposited, wash the residues from the planchette with distilled water, blot it dry with a paper tissue and finally dry it on a hotplate.

4. Place the prepared bismuth source on the top shelf of a Geiger–Müller counter castle and determine its activity at approximately 15 min intervals over a period of 2 h. The counting time for each determination should be long enough to record 10 000 counts.

5. Plot $\log_{10}$ (counting rate) for the bismuth source against the average time at which each counting rate was measured. Determine the half-life of bismuth-212 by the method described in Experiment 1.

Note. The 10·6 h half-life of the combined lead-212/bismuth-212 source may be determined before chemical separation by making activity measurements directly on the collecting foil over a period of about 2 days.

Experiment 21

Radiochemical separation of praseodymium-144 from its parent, cerium-144

Cerium-144 is a product of the fission of uranium-235 and is obtained by chemical separation from the used fuel rods of nuclear reactors. The decay sequence of cerium-44 is as follows:

$$^{144}_{58}\text{Ce} \xrightarrow{\beta} {}^{144}_{59}\text{Pr} \xrightarrow{\beta} {}^{144}_{60}\text{Nd} \text{ (stable neodymium-144)}$$

Half-life	285 d	17 m
β-particle	0·18 (24%)	3·0 (97%)
energies in MeV	0·31 (76%)	2·3 (1%)
		0·8 (2%)

After the separation of cerium-144 from a fission product mixture, secular equilibrium will be established with praseodymium-144 within about five half-lives of the praseodymium-144. The short-lived isotope may therefore be obtained by separation from cerium-144 preparations. A method for separating praseodymium from its parent is outlined in Fig. E.21.1. Cerium is separated as the slightly soluble ceric iodate, tervalent praseodymium forming a soluble iodate. The cerium is then redissolved as nitrate after reduction with hydrogen peroxide. This facilitates waste disposal, enabling the disposal to be made in solution rather than in solid form. It is also possible to use the recovered cerium-144 again, passing it through the oxidation cycle again after the addition of more praseodymium carrier. In practice, however, the quantities of activity used in an experiment are very small and they are not worth recovering.

The β-particles from praseodymium-144 are of high energy and may be detected using a thin glass-walled Geiger–Müller counter. The decay of the praseodymium-144 may therefore be followed if the filtrate from the ceric iodate precipitation is placed in a liquid counter. The parent, cerium-144, emits only low-energy β-particles which cannot penetrate the walls of such a counter. If the recovered cerium solution is placed in a liquid counter, the counting rate will therefore be low initially and will increase as the praseodymium-144 grows in the solution. Such observation of the growth of praseodymium, however, requires a rapid recovery of the iodate precipitate.

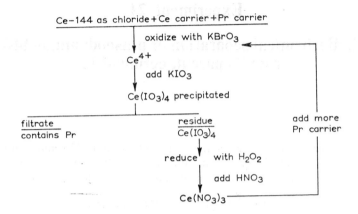

Fig E.21.1 Separation of praseodymium-144 from cerium-144

$6H^+ + BrO_3^- + 6Ce^{3+} \rightarrow Br^- + 6Ce^{4+} + 3H_2O$
$Ce^{4+} + 4IO_3^- \rightarrow Ce(IO_3)_4$
$2Ce^{4+} \text{ (as iodate)} + H_2O_2 \rightarrow 2Ce^{3+} + 2H^+ + O_2$

Experimental

(A) *Separation of praseodymium-144 from cerium-144 and the determination of its half-life*

Labelled material Cerium-144 labelled cerous chloride (Radiochemical Centre catalogue number CHS 1)
Recommended activity 0·1 µCi

1. Dissolve about 40 mg of cerous ammonium nitrate in 10 cm³ of 5M nitric acid in a 50 cm³ beaker. Add 3 cm³ of a praseodymium carrier solution containing about 1 mg of praseodymium/cm³ in 5M nitric acid. Add approximately 0·1 µCi of cerium-144/praseodymium-144 stock solution.

2. Add about 0·25 g of potassium bromate to the solution and warm gently to oxidize the cerium. Cool the solution to ice temperature.

3. After the bromate has completely dissolved, add slowly, with stirring, a solution of approximately 1 g of potassium iodate in a little distilled water. Stand the precipitate of ceric iodate in ice for about 5 min, stirring frequently. Add a large excess of iodate to depress the solubility of ceric iodate.

4. Filter in cascade through a Whatman No. 1, 41 or 541 filter paper, followed by a 42 or 542 filter paper, into a 50 cm³ beaker. Pour the filtrate into a liquid Geiger–Müller counter and determine the half-life of praseodymium-144 as described in Experiment 1.

Experiment 21

(B) *To observe the growth of praseodymium-144 in a ceric iodate, cerium-144-labelled precipitate*

After separation of praseodymium-144 from its parent cerium-144, the praseodymium activity will re-establish secular equilibrium with the parent in about 90 min. After one half-life of praseodymium (17 min) the activity will have reached one-half of its equilibrium value. It is therefore necessary to effect the chemical manipulations as rapidly as possible so that counting may be started as soon as possible after the instant of precipitation.

1. Dissolve about 10 mg of cerous ammonium nitrate in 2 cm^3 of 5M nitric acid in a centrifuge cone. Add 2 cm^3 of praseodymium carrier solution and about 0·1 μCi of cerium-144/praseodymium-144 stock solution.

2. Add about 0·1 g of potassium bromate to the solution and warm gently to effect oxidation of the cerium. Cool to room temperature.

3. Add just sufficient of a concentrated potassium iodate solution to effect complete precipitation of ceric iodate. Spin down the precipitate and wash it twice with about 5 cm^3 of a saturated solution of potassium iodate in 5M nitric acid. Wash once with distilled water and transfer a small portion of the precipitate to an aluminium planchette. Dry the source as rapidly as possible on a hotplate.

4. Place the source under an end-window Geiger–Müller counter and cover it with a 100 mg/cm^2 aluminium absorber to absorb the low-energy β-particles from the cerium-144. Determine the counting rate, due to praseodymium-144, at intervals over the next hour. For each count make the counting time sufficiently long to obtain good statistical accuracy, and record the actual time of the mid-point of each count. Correct each counting rate for counter dead time and background, and plot the corrected counting rate against time.

Note. Recovery of cerium-144 (if desired). To the precipitate in the filter funnel add slowly 10 cm^3 of 10 volumes hydrogen peroxide, followed by 5 cm^3 of 5M nitric acid. Repeat, if necessary, to dissolve all the precipitate.

REFERENCE

BRADLEY, A., and ADAMOWICZ, M. "Separation of cerium-144 from praseodymium-144," *Journal of Chemical Education*, **36,** 136 (1959)

Experiment 22

Preparation of iodine-128 using a laboratory neutron source and the Szilard-Chalmers reaction to concentrate the activity

The production of short-lived radioisotopes may be carried out with the use of neutron sources in cases where the cross-section for the nuclear reaction is high. The activity induced in N_{tg} target atoms of an element after irradiation for time t in a neutron flux ϕ is given by

$$\frac{dN_r}{dt} = \sigma N_{tg} \phi (1 - e^{-0.693 t/t_{1/2}})$$

where $t_{1/2}$ is the half-life of the radioisotope produced and σ is the activation cross-section for the element irradiated (see eqn. (2.9)). Thermal neutron fluxes of 10^{12} n/cm² s or greater can be obtained in nuclear reactors, and very many materials may be activated in them, but in the case of laboratory neutron sources the thermal neutron flux is only of the order of 10^2 to 10^4 n/cm² s and consequently only elements with high cross-sections may be activated. The saturation factor $(1 - e^{-0.693 t/t_{1/2}})$ reaches a value of 0·5 for $t = t_{1/2}$ and 0·97 for $t = 5 t_{1/2}$, so that a high percentage of the maximum activity may be obtained by irradiation for between one and five half-lives of the isotope being produced. It is possible to obtain useful activities for only a few isotopes by irradiation in a neutron source, for example for

^{127}I (n, γ) ^{128}I $t_{1/2} = 25$ min $\sigma = 5\cdot 6$ barns

and

^{55}Mn (n, γ) ^{56}Mn $t_{1/2} = 2\cdot 6$ h $\sigma = 13\cdot 3$ barns

Neutron sources produce neutrons by the interaction of α-particles or γ-photons on beryllium nuclei. Such sources are prepared by sealing a mixture of an α-particle or γ-ray emitting isotope and beryllium in a platinum or aluminium capsule. The α-emitting isotopes used include radium-226 and americium-241, whilst antimony-124 is commonly used for the γ-neutron sources. The characteristics of these sources are given in Table 27.

The use of a small neutron source (about 10 mCi) may not produce sufficient activity in a sample to be directly measurable. Increasing the sample size (i.e. N_{tg}) will increase the total activity produced, but

Experiment 22

Table 27 Neutron sources

Neutron source	Neutron production reaction	Half-life	Neutron output, n/s for 1 Ci	γ-ray dose at 1 m, R/h per Ci
Ra/Be	$^9_4\text{Be}(\alpha, n)^{12}_6\text{C}$	1620 y	$1 \cdot 3 \times 10^7$	0·825
Am/Be	$^9_4\text{Be}(\alpha, n)^{12}_6\text{C}$	458 y	$2 \cdot 5 \times 10^6$	0·025
Sb/Be	$^9_4\text{Be}(\gamma, n)^8_4\text{Be} \to 2^4_2\text{He}$	60 d	$1 \cdot 3 \times 10^6$	0·98

it then becomes necessary to concentrate the active species in some way so that suitable counting geometries can be obtained. This can sometimes be accomplished by means of the Szilard–Chalmers reaction, the theory of which is outlined below.

SZILARD–CHALMERS REACTIONS

After a neutron capture event the target nucleus is left in an excited state and it radiates this excitation energy as a complex spectrum of γ-rays of several MeV energy. The γ-rays are emitted within 10^{-20} to 10^{-12} s after the capture event, and the reaction is therefore referred to as an (n, γ) reaction. The process is thus:

$$^A_Z\text{X} + ^1_0\text{n} \to {}^{A+1}_Z\text{X} \text{ excited nucleus}$$
$$\downarrow 10^{-23} \text{ to } 10^{-12} \text{ s}$$
$$^{A+1}_Z\text{X} + \gamma\text{s of several MeV}$$

The resulting nucleus may be either stable or radioactive. When a nucleus emits a γ-ray, which may be either a capture γ-ray or a γ-ray emitted in the process of radioactive decay, a recoil momentum is imparted to the nucleus. Considering an atom of mass m, which emits a γ-ray, conservation of momentum requires that

Momentum of γ-ray = Momentum of recoil atom

so that the γ-ray momentum is

$$p = \frac{h\nu}{c} = \frac{E_\gamma}{c} = mv$$

where E_γ is the energy of the emitted γ-photon, and v is the velocity of the recoil nucleus of mass m.

The kinetic energy of the recoiling nucleus is therefore given by

$$E_m = \tfrac{1}{2}mv^2 = \frac{p^2}{2m}$$
$$= \frac{E_\gamma^2}{1862m}$$

where E_γ is measured in mega-electronvolts and m in atomic mass units.

In the case of a molecule, this recoil energy will subsequently be partitioned between the atoms of the molecule, appearing both as kinetic energy and as excitation energy which may dissociate the molecule. For a molecule of total mass $(m + m')$, containing an atom of mass m which emits a γ-photon as described above, conservation of momentum requires that

$$p = (m + m')v$$

But the kinetic energy of the molecule, $E_{(m+m')}$, is given by

$$E_{(m+m')} = \tfrac{1}{2}(m + m')v^2$$

and eliminating the velocity, v, from these equations,

$$p^2 = 2(m + m')E_{(m+m')}$$

When the γ-ray is emitted and the recoil energy of the nucleus is partitioned as molecular kinetic energy and excitation energy, conservation of energy requires that

$$\text{Recoil energy of nucleus} = \text{Kinetic energy of molecule} + \text{Excitation energy of molecule}$$

so that the excitation energy is given by

$$E_{\text{excitation}} = \frac{p^2}{2m} - \frac{p^2}{2(m+m')}$$

$$= \frac{p^2}{2m} \frac{m'}{m+m'}$$

$$= \frac{E_\gamma^2}{1862m} \frac{m'}{m+m'} \text{ mega-electronvolts}$$

Table 28 gives values of E_γ for neutron capture by the halogens, the excitation energy of hydrogen halides and alkyl halides calculated as above, and the corresponding bond energies. In most of

Table 28 Gamma-ray recoil energies for neutron capture and the Szilard–Chalmers effect in halides

Halogen	Energy of most abundant capture γ-ray, MeV	Recoil energy of nucleus, eV	C_2H_5—X		H—X	
			Bond energy, eV	Excitation energy, eV	Bond energy, eV	Excitation energy, eV
Chlorine-36	6·11	530	2·89	229	4·46	13·5
Bromine-80	7·56	380	2·35	101	3·79	4·7
Iodine-128	7·0	210	1·98	39	3·10	1·6

Experiment 22 205

the cases quoted, the molecular excitation energy due to γ-recoil is great enough to break the chemical bond involved. This breaking of chemical bonds following neutron capture is known as the *Szilard–Chalmers reaction*. With ethyl iodide the reaction is

Nuclear reaction $^{127}I - C_2H_5 + {}_0^1n \rightarrow ({}^{128}I - C_2H_5)^*$ ⌐
Szilard-Chalmers reaction $^{128}I^* + \cdot C_2H_5 + \gamma$-photons ⌐

The precise fate of the energetic free radicals formed is complex and depends upon the nature of their surroundings. Thus the mole fractions of any solvent/solute species in the system can influence their behaviour. If radical scavengers such as molecular iodine are present then a large fraction of the radioactive ^{128}I atoms liberated by the Szilard–Chalmers effect may be recovered as labelled iodine.

Experimental

A suitable neutron source is a 10 mCi radium/beryllium or americium/ beryllium source. This should be mounted at the end of a handling rod (length about 60 cm) so that the dose rate at the handle end, due to both neutrons and γ-rays is at a safe level. The source end of the rod must always be directed away from the person, and the source must be kept in a shielded container when not in use.

1. Place 500 cm³ of ethyl iodide in a round flask and dissolve about 0·1 g of iodine in it.

2. Place the neutron source in the ethyl iodide so that the source is at the centre of the flask, and irradiate the ethyl iodide for 2–3 h. Clamp the source in position by means of its handling rod and shield the flask during irradiation.

3. After activation, remove the neutron source, return it to its storage container and pour the ethyl iodide into a large separating funnel. Add 10 cm³ of a solution containing 1 % sodium thiosulphate and 1 % potassium iodide and shake for about 5 min. This extracts both iodide and free iodine from the ethyl iodide.

4. Place 10 cm³ of the ethyl iodide in a liquid-sample Geiger–Müller counter and determine its activity. Correct this activity for the counter background, and then calculate the total iodine-128 activity remaining in the 500 cm³ of ethyl iodide.

5. Recover the aqueous extract and determine its activity using the same Geiger–Müller counter. Determine the half-life of the iodine-128 by the method described in Experiment 1.

6. Calculate the percentage of the iodine-128 activity which has been extracted from the ethyl iodide. For this purpose, the activities of the ethyl iodide and the extract should be corrected to the time of extraction.

Notes

1. The initial activity to be expected when measured with a liquid-sample Geiger–Müller counter is about 5000 counts per minute for the extract. This is about 75% of the total activity produced.

2. The hazards associated even with small neutron sources, such as that used for this experiment, are much greater than those arising from the use of the radioactive materials for the other experiments described in this book. It is not possible for an establishment to carry out work involving a neutron source within the provisions of the Radioactive Substances Act, Schools Exemption Order, 1963, and a special authorization must therefore be obtained before a neutron source can be purchased.

Experiment 23

The use of nuclear emulsions for the detection of alpha particles

Nuclear emulsion plates may be used as α-particle detectors. If an α-particle loses its kinetic energy within the photographic emulsion of such a plate, the silver halide grains along the track of the α-particle become sensitized and may be rendered visible by reduction of the grains to metallic silver in the photographic development process. The α-particle tracks produced in a nuclear emulsion by uranium decay are shown in Plate 12. Since it is necessary to ensure that the α-particles are entirely absorbed within the emulsion, the material whose α-activity is being investigated must either be in direct contact with the emulsion surface or be absorbed within it. In the case of solutions of α-particle emitters there are two methods by which the radioactive material may be incorporated in the emulsion. Either a small aliquot of the solution may be evaporated on the surface of the emulsion or the plate may be immersed in the solution so that the emulsion will absorb some of the radioactive solute. In the first method the quantity of the radionuclide deposited on the surface of the emulsion may be calculated from the droplet size and the concentration of the solution. If the plate is totally immersed in a radioactive solution, however, the amount of the radionuclide absorbed by the emulsion is not directly proportional to the concentration of the solution. To determine the amount of the radionuclide absorbed by the emulsion it is therefore necessary to chemically analyse the solution after the plate has been removed.

The thickness of a nuclear emulsion used for α-particle detection should be somewhat greater than the range of the α-particles in it, but not so great that observations become difficult due to a multiplicity of tracks at different depths in the emulsion. A thickness of about 50 μm is generally used for α-particle detection. These emulsions require great care in handling at all stages of processing to prevent mechanical damage. They also require longer developing, fixing and washing times than normal photographic materials because of their relatively great thickness.

The number of tracks in a 1 cm^2 area of a nuclear emulsion plate is known as the *track density*. This will be a function both of the concentration of the radionuclide within or on the surface of the emulsion and of the exposure time. The exposure time is the time which elapses between loading the emulsion with the radionuclide and developing the plate. It is the time during which decay of the radionuclide will be recorded as tracks in the emulsion. With an exposure of 3 to 4 days it is possible to detect about 10 μg of uranium deposited by evaporation on the emulsion. The distribution of uranium atoms in the surface of a rock section containing about 1 part per million of uranium may also be determined, but this requires an exposure of the order of 2 to 3 months.

EXPERIMENT 23.1 A STUDY OF THE RELATIONSHIP BETWEEN URANIUM CONCENTRATION AND ALPHA-PARTICLE TRACK DENSITY

The first few members of the uranium series are shown in Table 25, and the complete series will be found in Appendix 3. If a long-lived radioactive isotope decays to a relatively short-lived daughter then radioactive equilibrium between the parent and daughter isotopes will be established after a few half-lives of the daughter. Before a mineral containing uranium is chemically processed, uranium-238 and all its daughter products are generally in radioactive equilibrium. When uranium is chemically separated from the mineral, uranium-238 and uranium-234 are the only isotopes present (neglecting the small contribution from uranium-235). These isotopes will be present in the equilibrium ratio of about 19 000 atoms of uranium-238 to each atom of uranium-234, but it should be noted that their decay rates will be identical. A few months after chemical separation of uranium, radioactive equilibrium will be completely re-established as far as uranium-234 in the decay series, but it would require millions of years to re-establish equilibrium throughout the series. Thus the only α-emitting isotopes present in commercial uranium salts are uranium-238 and uranium-234. The α-particle activity of a solution of a uranium salt is therefore related to the uranium concentration.

Experimental

Materials

   Nuclear emulsion: Ilford type K2 nuclear research plates, either 3 in × 1 in or 1 in × 1 in with an emulsion thickness of 50 μm

Developer: Kodak type D19(b). Dissolve in the order given:

Metol	2·2 g
Sodium sulphite (crystalline)	144·0 g
Hydroquinone	8·8 g
Sodium carbonate (crystalline)	130·0 g
Potassium bromide	4·0 g
Distilled water	1000 cm^3

Fixer: A filtered solution containing 30 g of sodium thiosulphate and 2·25 g of anhydrous sodium bisulphite in 100 cm^3 of distilled water.

Safelight: Ilford S or Kodak 6B (or 6BR)

1. Prepare solutions containing 0·25, 0·5, 1·0, 1·5 and 2·0 g of uranyl acetate, $UO_2(CH_3COO)_2 \cdot 2H_2O$, in 50 cm^3 of distilled water. These solutions contain 28, 56, 112, 168 and 224 μg of uranium in 0·01 cm^3 respectively.

2. Use an Agla syringe to place several 0·01 cm^3 droplets of each of these solutions on the surface of nuclear emulsion plates. The geometry of all droplets should be kept as nearly as possible the same.

3. Place the prepared plates in a desiccator to evaporate the solution and then store them in total darkness for between 3 and 7 days. Develop the plates by the following method:
   (i) Immerse for 10 min in distilled water.
   (ii) Develop for 15 min in filtered developer.
   (iii) Wash for 10 min in running water.
   (iv) Fix in filtered fixing solution for 50% longer than the clearing time (about $\frac{3}{4}$ h).
   (v) Wash for 2 h in running water and then for 10 min in distilled water.
   (vi) Dry in a desiccator.

4. Examine the plates using a microscope with a magnification of about ×400. Use a squared eyepiece graticule and count the number of α-particle tracks that occur in unit area for each evaporated droplet of uranium solution.

5. Plot the number of tracks per unit area for each evaporated droplet against its uranium content.

EXPERIMENT 23.2 DETERMINATION OF THE HALF-LIFE OF URANIUM-238 USING A NUCLEAR EMULSION PLATE FOR THE DETECTION OF ALPHA PARTICLES

The only α-emitting isotopes present in uranyl salts are uranium-238, uranium-234 and uranium-235. Uranium-235 normally contributes about 4% of the total α-disintegration rate of natural uranium, but much of the material now supplied by the chemical manufacturers is deficient in uranium-235. Uranium-238 and uranium-234 will be present in their secular equilibrium amounts, so that the total α-disintegration rate will be twice that of uranium-238 (neglecting any contribution from uranium-235). In Experiment 1, a method for determining the half-lives of long-lived isotopes was described, which gave

$$\lambda = -\frac{\Delta N}{\Delta t}\bigg/N$$

where ΔN is the number of disintegrations occurring in the finite time interval Δt, and N is the total number of radioactive atoms present. The quantity N must be determined absolutely, and for this purpose a quantity of a uranium salt is incorporated in a nuclear emulsion. The emulsion will record the tracks of all the α-particles emitted within it up until the time of development. The number of uranium atoms incorporated in the emulsion can be determined by chemical analysis of the solution in which the plate was immersed.

Experimental

1. Prepare a standard uranium solution by dissolving 0·0445 g of uranyl acetate, $UO_2(CH_3COO)_2.2H_2O$, in 250 cm³ distilled water in a volumetric flask.

2. Prepare a trough of folded paper or thin card which closely fits a 3 in × 1 in nuclear emulsion plate and is about 1 cm deep. Dip the trough in molten paraffin wax to render it waterproof

3. Pipette 10 cm³ of the standard uranium solution into the paper tray, and immerse a nuclear emulsion plate in it so that the emulsion side is uppermost. Make sure that the emulsion is wetted by the solution, and leave it to soak for 30 min.

4. Remove the plate from the trough, wash any surplus solution back into the trough with a jet of water from a wash-bottle. Quantitatively transfer the solution from the trough to a 50 cm³ volumetric flask.

5. Place the nuclear plate in a light-tight box together with a small beaker containing silica gel. Leave the plate in this box for about a week to allow uranium decay to occur. Record as accurately as possible the time from the removal of the plate from the uranium solution until it is developed. This is the uranium decay time, during which all uranium decays will have been recorded in the emulsion.

6. Determine the uranium remaining in the solution in which the nuclear plate was imersed by the following method. Pipette 1·0, 2·5, 5·0 and 10·0 cm³ of the standard uranium solution into 50 cm³ volumetric flasks. Treat these solutions and the solution from the

paper trough as follows. Add 5 cm³ of M hydrochloric acid and 5 cm³ of 30% ammonium thiocyanate solution, make up to volume with distilled water, and leave the solutions to stand for 15 min with occasional shaking.

7. Measure the optical absorbance of the solutions against a reagent blank at 375 nm, using a spectrophotometer. Plot optical absorbance against the uranium content of the solutions, and so determine the uranium content of the solution in which the plate was immersed. The standard uranium solution contains 1 mg of uranium in 10 cm³. The uranium absorbed by the nuclear emulsion can now be found by difference.

8. At the end of the exposure period, process the nuclear emulsion plate by the method described previously. Inspect the processed plate using a microscope with a magnification of ×400. Determine the track density for several regions of the plate. Calculate the total number of tracks recorded on the plate from the plate area and the track density.

9. Calculate the half-life of uranium-238 from the number of uranium decays recorded on the plate, the decay time and the quantity of uranium absorbed in the emulsion.

EXPERIMENT 23.3 THE OBSERVATION OF THORIUM STARS IN A NUCLEAR EMULSION

There are two isotopes of thorium which occur in the thorium decay series. Thorium-232 has a half-life of $1 \cdot 39 \times 10^{10}$ years, and thorium-228 has a half-life of 1·9 years. After thorium-228 there are no isotopes in the series with half-lives greater than a few days. Although it would require some 30 to 40 years for secular equilibrium to be re-established throughout the series after chemical separation of the thorium, equilibrium from thorium-228 onwards will be established in about 20 days. Provided that several days have elapsed since the thorium was chemically separated from its ore, there will be appreciable quantities of the daughter products of thorium-228 present. The main decay sequence from thorium-228 onwards is given in Table 26.

Thorium X (radium) may be separated from a thorium compound using a barium carrier, because of the similar chemical properties of radium and barium. If photographic plates are exposed to such a radium solution then the successive α-particles emitted in the decay chain will make a four-pointed track in the photographic emulsion as shown in Plate 13. Bismuth, however, decays by two modes with the same half-life, but the β, α decay leads to the emission of an α-particle whose energy is 20–25% greater than that of any other α-particle of the decay chain. Thus the star resulting from decay by this route will have one exceptionally long track. There should be approximately twice as many stars showing this exceptionally long track as have four tracks of nearly equal length. Stars with less than four tracks are due to the fact that development of the photographic plate has taken place before these particular atoms have completely decayed. Nuclear emulsions which are exposed to thorium solutions will show stars containing from one to five tracks corresponding to decay from thorium-228 onwards. Some thorium stars are shown in Plate 13.

Experimental

1. Dissolve 0·4 g of thorium nitrate and 5–10 mg of barium nitrate in the minimum amount of distilled water (about 10 cm^3). Add 10 cm^3 of 50% ammonium hydroxide solution, boil for a few minutes, and filter off the thorium hydroxide precipitate using a No. 3 porosity sintered glass crucible. The filtrate contains the soluble hydroxides of barium and radium. Boil the filtrate gently to expel excess ammonia, make it faintly acid with dilute nitric acid, and then adjust the volume to about 25 cm^3.

2. Immerse 1 in × 1 in pieces of nuclear emulsion plate in this solution for about 30 min, taking care not to damage the emulsion. Remove the plates from the solution, rinse rapidly in distilled water, and then dry them in a desiccator.

3. Store the dried plates in the dark and develop them after about 1, 3, 6, 9 and 12 days by the method described above.

4. Examine the plates with a microscope at a magnification of about 200 times, and determine the star density (irrespective of the number of tracks in the star) for each.

5. Plot the number of stars per unit area against the decay time, and from this plot determine the half-life of radium-224.

6. The branching ratio of bismuth-212 may be determined by estimating the relative number of stars which show one exceptionally long track. A magnification of 400 to 600 times should be used, and only four pointed stars are considered. Stars for which a definite decision cannot be made should be ignored, but the accuracy of the result will depend upon the number of decisions made.

Note. If the developed plates from these experiments are to be kept permanently, they should be covered with glass cover slips using DPX mountant.

Experiment 24

Determination of the specific surface area of a lead sulphate precipitate using sulphur-35-labelled sulphate

If a tracer amount of sulphur-35 is added to a saturated solution of lead sulphate, an equilibrium is established between the labelled species on the surface of the precipitate and in the solution, as follows:

SO_4 (surface) + $^{35}SO_4$ (solution)
$$\rightleftharpoons {}^{35}SO_4 \text{ (surface)} + SO_4 \text{ (solution)}$$

and at equilibrium,

$$\frac{[^{35}SO_4]_{surface}}{[^{35}SO_4]_{solution}} = \frac{[SO_4]_{surface}}{[SO_4]_{solution}}$$

The concentrations of the labelled species may be measured radiometrically, and the total sulphate concentration in the solution may be calculated from the solubility of lead sulphate. The quantity of sulphate in the surface of the lead sulphate precipitate may then be calculated.

The exchange will take place only between sulphate on the surface of the precipitate and the sulphate in solution, as the self-diffusion of sulphate ions into the interior of the lead sulphate particles occurs at a relatively insignificant rate. The method may be used to study the processes involved in ageing and coagulation of precipitates.

Experimental

1. Weigh accurately about 0·25 g of lead sulphate into a 25 cm^3 volumetric flask, and add distilled water to make the total volume 25 cm^3. Allow the solution to stand for several days with occasional shaking to obtain a saturated solution of lead sulphate.

2. Add 1 μCi of carrier-free sulphur-35 labelled sulphate solution to the volumetric flask containing the saturated solution of lead sulphate. The specific activity of the labelled sulphate solution should be such that a negligible volume, about 0·01 cm^3, of solution is added to the saturated solution. Shake the volumetric flask vigorously for 1 h in a mechanical shaker to permit equilibrium for the labelled/unlabelled sulphate exchange to take place.

3. Transfer the contents of the volumetric flask immediately to a 50 cm^3 centrifuge cone, spin down the precipitate, and transfer the

saturated solution to a second centrifuge cone. Rinse the flask with about 10 cm³ of ethyl alcohol, add this wash to the centrifuge cone containing the precipitate, spin down and transfer the alcohol to the cone containing the saturated solution. The first centrifuge cone now contains all the lead sulphate precipitate, and the second the total activity which remained in the saturated lead sulphate solution.

4. Quantitatively transfer the lead sulphate precipitate, as a slurry with alcohol, to an aluminium counting planchette, and prepare as evenly distributed a source as possible for counting, evaporating the slurry to dryness under an infra-red lamp.

5. To the solution in the second centrifuge cone add precisely the volume of 0·1M sulphuric acid to make the sulphate present exactly equivalent to that present as lead sulphate originally. Add a slight excess of 1M lead acetate, stirring vigorously, and then allow the solution to stand to coagulate the colloidal precipitate. Spin down the precipitate and prepare a source as described above. This source will contain exactly the same weight of lead sulphate as the first source, and the counting rates of the two sources may be compared directly without making any corrections for self-absorption of the weak sulphur-35 β-particles.

6. Determine the counting rates of the two sources with an end-window Geiger–Müller counter, using the same counter geometry for each. Correct the counting rates for paralysis time and background. These counting rates are directly proportional to the concentrations of $^{35}SO_4^{2-}$ on the surface of the lead sulphate precipitate and in the solution.

7. Calculate the mass of SO_4^{2-} present in the precipitate as surface sulphate and hence the specific surface area:

Specific = Number of molecules in = Surface area for
surface   surface/per gramme   a molecule
area   of precipitate

Assuming that the molecules have a cubic symmetry the area per molecule is given by

$$\left[\frac{(\text{Mol. wt.})}{(6\cdot02 \times 10^{23})\rho}\right]^{2/3}$$

where ρ, the density of lead sulphate, $PbSO_4$, is 6·23 g/cm³. The solubility of lead sulphate in water is 0·0425 g/1000 cm³ at 25°C.

Further work. The variation of the surface area of freshly precipitated lead sulphate and lead sulphate precipitates aged in various ways may be investigated.

REFERENCE

NEALY, C. L. and CHOPPIN, G. R.   *Journal of Chemical Education*, **11**, p. 597 (1964)

Experiment 25

Determination of the partition coefficient for iodine between carbon tetrachloride and water

The partition coefficient for the distribution of a solute between two immiscible solvents is defined as

$$k = C_1/C_2$$

where C_1 and C_2 are the concentrations of the solute in each solvent, for cases where the molecular weight, or degree of association, of the solute is the same in both solvents. These concentrations may be very readily determined by radiometric methods if the solute is labelled with a radioisotope. This method is of wide applicability and is much used, for example, in analytical chemistry to determine the efficiency of solvent extraction procedures. The method may be illustrated by an investigation of the distribution of iodine between water and carbon tetrachloride; the solubility of iodine in water may also be readily determined.

Experimental

Labelled material   Iodine-131-labelled sodium iodine (Radiochemical Centre catalogue number IBS 1)

Recommended activity 1·0 μCi

(A) *Preparation of iodine-131-labelled iodine in carbon tetrachloride solution*

1. Prepare 25 cm³ of an M/10 solution of potassium iodide, adding about 1·0 μCi of carrier-free iodine-131-labelled sodium iodide solution before adjusting the volume.

2. Place this solution in a separating funnel containing 15 cm³ of carbon tetrachloride. Dissolve 0·8 g of potassium dichromate in the aqueous layer, and then add 1·0 cm³ of concentrated sulphuric acid. Extract the liberated iodine into the carbon tetrachloride layer. (Carry out operations involving free iodine in a fume cupboard.) Separate the carbon tetrachloride layer and wash it with two 20 cm³ aliquots of distilled water.

215

(B) *Determination of the partition coefficient*

1. Pipette 10 cm^3 of the carbon tetrachloride solution of iodine and 10 cm^3 of distilled water into a separating funnel and shake for 10–15 min. Allow the layers to separate.

2. Carefully run the carbon tetrachloride layer into a liquid Geiger–Müller counter and then determine the counting rate with good statistical accuracy. Empty, wash and dry the counter and check its background, then run the aqueous layer into it and determine its activity with good statistical accuracy. The two counting rates are proportional to the concentrations of iodine in each phase, and their ratio gives the partition coefficient.

(C) *The equilibrium* $I^- + I_2 \rightleftharpoons I_3^-$

The dynamic nature of this equilibrium may be demonstrated using an iodine-131-labelled iodide solution.

1. Add 1·0 μCi of carrier-free iodine-131-labelled sodium iodide solution to 20 cm^3 of M/10 potassium iodide solution.

2. Show that no activity can be extracted from this solution by 10 cm^3 of carbon tetrachloride.

3. Prove that activity is extracted by 10 cm^3 of an M/10 solution of iodine in carbon tetrachloride. Show that radioactive iodine may now be extracted from the remaining aqueous phase by 10 cm^3 of carbon tetrachloride.

REFERENCE

PEACOCKE, T. A. H.   "Some applications of radioisotope techniques in chemistry and biology,"   *School Science Review*, **45**, p. 597 (1963–64)

Experiment 26

Radiometric analysis

The decay rate of a radioactive element is related to the quantity of the element present by the radioactive decay law:

$$\frac{dN}{dt} = -\lambda N$$

If the radioactive decay constant is known and the absolute disintegration rate of a sample can be determined, it is thus possible to calculate the mass of the element present. Many radioisotopes can, in fact, be determined with great sensitivity by this method. Radium, for example, can be determined in rocks, soils or drinking water in amounts corresponding to as little as 10^{-12} g. In the determination of elements which are not themselves radioactive, the radiometric method can be applied if it is possible to form a compound with a radioactive reagent and subsequently separate the excess reagent from the compound formed. The radioactivity of the labelled compound so formed will be related to the quantity of the element from which it was formed and the specific activity of the reagent. This quantity of the element present may be calculated from the measured activity of the compound and the specific activity of the reagent, provided that the stoichiometry of the reaction is known.

EXPERIMENT 26.1 RADIOMETRIC DETERMINATION OF MAGNESIUM

Magnesium may be determined radiometrically by precipitation of magnesium phosphate, $Mg_3(PO_4)_2 \cdot xH_2O$, with a solution of phosphorus-32-labelled sodium phosphate and subsequent recovery and measurement of the activity of the precipitate. In ammoniacal solutions at pH 11–12, magnesium may readily be precipitated as magnesium phosphate. At higher concentrations of alkali, magnesium hydroxide is precipitated. The precipitate must be recovered quantitatively, and for quantities of magnesium in the range 1–10 mg this recovery can be performed without the use of a carrier precipitate. For smaller amounts of magnesium, it is necessary to use a carrier precipitate, and a preformed precipitate of magnesium ammonium phosphate may be used since exchange between excess labelled phosphate and phosphate ions in this precipitate does not take place.

Since two moles of phosphate are required to precipitate three moles of magnesium, if S is the specific activity of the labelled phosphate in counts per minute per mg of PO_4^{3-} ions, and A is the activity of the magnesium phosphate precipitate in counts per minute, the mass of magnesium precipitated, w, is

$$w = \frac{A(3 \times 24\cdot32)}{S(2 \times 95\cdot02)} \text{ mg}$$

Experimental

Labelled material — Phosphorus-32-labelled sodium orthophosphate (Radiochemical Centre catalogue number: PBS 1)
Recommended activity 1·0 to 10·0 µCi

1. Prepare 100 cm³ of a solution of magnesium sulphate containing 5 mg/cm³ of Mg^{++}. Use this solution to prepare a more dilute solution containing 0·5 mg/cm³ of Mg^{++}. These solutions may be used to dispense standard amounts of magnesium from 0·05 mg of Mg^{++} upwards.

2. Prepare 100 cm³ of a solution containing about 1 mg/cm³ of PO_4^{3-} and a total activity of 1 µCi of phosphorus-32-labelled phosphate. Prepare a similar solution containing a total phosphorus-32 activity of 10 µCi. The solution of higher specific activity should be used for determinations of quantities of magnesium below 1 mg.

3. Make an investigation of the accuracy of the method for the determination of magnesium in the range 50 µg of Mg^{++} to 10 mg of Mg^{++}. Carry out the precipitations as described below.

In the range 1–10 mg *of* Mg^{++}

To the magnesium solution in a small beaker or test tube add an excess of the labelled phosphate solution. Add a few drops of 0·01M ammonium hydroxide to precipitate magnesium phosphate. Recover the precipitate on a small filter paper disc in a demountable filter and wash it with 2M ammonium hydroxide and then with alcohol.

In the range 50–1000 µg *of* Mg^{++}

To a solution containing about 2 mg of Mg^{++} ions in a centrifuge cone, add an excess of 0·5M di-ammonium hydrogen phosphate followed by 0·1M ammonium hydroxide to precipitate magnesium ammonium phosphate. Spin down the precipitate and wash it with distilled water. Add the magnesium solution to be determined to the precipitate in the centrifuge cone, and then add an excess of the

Experiment 26.2

labelled phosphate solution of higher specific activity. Precipitate magnesium phosphate by the addition of a few drops of 0·1M ammonium hydroxide. Recover and wash the combined precipitates on a demountable filter as before.

4. Transfer the precipitates on the filter paper discs to counting planchettes and dry them under an infra-red lamp. Determine their activities with an end-window Geiger–Müller counter.

5. Determine the activities of suitable aliquots of each of the labelled phosphate solutions. Evaporate the solution to dryness on a counting planchette and count them under the same conditions of counter geometry as used above.

6. Calculate the magnesium content of each precipitate.

Reference
MARTELLY, J., and SUE, P. *Bull Soc Chim France*, p. 103 (1946)

EXPERIMENT 26.2 RADIOMETRIC DETERMINATION OF LEAD

The radiometric determination of traces of heavy metals in biological materials has been described (Reference 1) using the labelled phosphate method. This experiment illustrates the principles of the method as applied to the determination of lead, using lead nitrate solutions of known concentration. A small aliquot of the solution in which lead is to be determined is evaporated on a strip of chromatography paper. A solution containing a phosphorus-32-labelled phosphate reagent is then added to the paper over the area where the lead is deposited. After evaporation of the reagent, the excess reagent is eluted away from the lead phosphate so formed by a chromatographic method.

The same method is then applied to the determination of lead impurity in nickel sulphate as an illustration of the kind of problem to which the radiometric method may in practice be applied. Since the quantity of nickel involved in this determination is greatly in excess of its lead impurity content, it is necessary to separate the lead from the nickel before formation of the labelled compound, and this is effected by a solvent extraction procedure.

Experimental: The radiometric determination of lead in the range 10 to 400 μg of lead
Labelled material   As Experiment 26.1
Recommended activity 10·0 μCi

1. Prepare a standard lead solution containing 0·4 g of lead nitrate $Pb(NO_3)_2$, in 100 cm^3 of distilled water. This solution will contain 25 μg of Pb^{++} in each 0·01 cm^3.

2. Prepare an inactive phosphate solution containing 2·5 g of disodium hydrogen phosphate, $Na_2HPO_4.12H_2O$, in 100 cm³ distilled water.

Pipette 1·0 cm³ of this solution into a small beaker or weighing bottle and add 10 μCi of a carrier-free phosphorus-32-labelled orthophosphate solution in a volume of about 1·0 cm³. The labelled solution so prepared is sufficient for about 40 determinations; 0·05 cm³ of it are equivalent to about 500 μg of Pb^{++}.

3. Cut several 2·5 cm × 20 cm pieces of Whatman 3MM chromatography paper and mark each with a small dart cut about 3 cm from one end. Using an Agla syringe, place 0·005, 0·01, 0·02 cm³, etc., respectively of the standard lead solution opposite the marks on the chromatograms. Dry the chromatograms with a hair-dryer.

4. To each of the prepared chromatograms add 0·5 cm³ of the labelled phosphate solution, adding the solution to the region where the lead has been deposited. Make this addition in two or three aliquots, drying the chromatogram between additions and finally dry the chromatogram as before.

5. Place the chromatograms in a chromatography tank and separate the excess phosphate reagent from the lead phosphate formed by ascending chromatography for about 1½ hours using a borax/oxalate buffer as eluant.

Borax/oxalate buffer. Mix 60 cm³ of ethanol, 60 cm³ of n-propanol, 20 cm³ of distilled water, 20 cm³ of 0·05M oxalic acid solution and 40 cm³ of 0·05M borax solution adjusting to pH 8·0 with the borax solution.

6. Dry the chromatograms and place the one with the highest lead content in a simple chromatogram scanner (Plate 3). Measure the activity due to phosphorus-32 at 0·5 cm intervals over the length of the chromatogram. An activity peak will be found due to lead phosphate near the original reference mark, and the excess labelled phosphate should have moved away to close behind the solvent front. Integrate the activities in the region of the lead phosphate. Similarly determine the total activities due to labelled lead phosphate on the other chromatograms.

7. Plot the total activities (corrected for counter dead time and background) due to lead phosphate on the chromatograms against the known lead contents.

Experiment 26.2 221

Further work: The determination of the lead content of AnalaR nickel sulphate

1. Prepare a 0·01% solution of diphenyldithiocarbazone (dithizone) in carbon tetrachloride. 1 cm^3 of this solution is equivalent to about 40 μg of lead.
2. Prepare an ammonia/cyanide/sulphite masking solution as follows. Dissolve 0·6 g of potassium cyanide in 30 cm^3 of distilled water and add it to 70 cm^3 of concentrated ammonium hydroxide solution. Dissolve 0·75 g of sodium sulphite in this solution (*caution*: this is poisonous).
3. Dissolve 5 g of AnalaR nickel sulphate in about 20 cm^3 of distilled water and transfer the solution to a separating funnel. Add 25 cm^3 of the masking solution and sufficient dithizone solution to contain a 50% excess of dithizone over the quantity required to extract the lead. (AnalaR nickel sulphate is quoted as containing up to 40 parts per million of lead.) Add sufficient carbon tetrachloride to make the volume of the organic layer up to about 25 cm^3. Shake the separating funnel for about 2 min and then allow the phases to separate.
4. Recover the carbon tetrachloride layer in a small beaker and then evaporate it to dryness under an infra-red lamp, placed in a fume cupboard. Treat the residues with two successive 2 cm^3 aliquots of concentrated nitric acid, evaporating each to dryness on a sand bath.
5. Weigh the beaker, dissolve the residues in about 0·25 cm^3 of distilled water and weigh again to determine the weight of solution. Transfer as much as possible of the solution to a chromatogram strip, making small additions and evaporating after each so that the lead ion is confined to a small region of the filter paper. Weigh the beaker containing the remainder of the solution and so determine the fraction which has been transferred to the chromatogram.
6. Form labelled lead phosphate on the chromatogram as described in the first part of the experiment. Determine the activity of the lead phosphate formed and use the standards of the first part of the experiment to determine the lead on the chromatogram. Calculate the lead content of the nickel sulphate in parts per million.
7. To determine the accuracy of the determination repeat the experiment using nickel sulphate solutions to which 50 and 100 μg of lead have been deliberately added. The lead contents found with these solutions should be 50 and 100 μg higher, respectively, than found for the nickel sulphate.

References

1 VAN ERKELENS, P. C. "Radiometric trace analysis of lead with phosphate-^{32}P", *Analytica Chimica Acta*, **25**, p. 570 (1961)
2 MORRISON, G. H., and FREISER, H. *Solvent Extraction in Analytical Chemistry* (Wiley 1957)

EXPERIMENT 26.3 RADIOMETRIC DETERMINATION OF COPPER IN A COPPER/NICKEL SOLUTION

Metals which form insoluble phosphates can, in principle, be determined by the formation of a labelled phosphate precipitate using a phosphorus-32-labelled phosphate reagent. After separation of the excess reagent, the radioactivity of the precipitate may be determined and is a measure of the quantity of the metal ion precipitated. The method can be applied at trace levels by using the technique of paper chromatography to separate the metal ions present and the excess reagent from the phosphate precipitate. Details for the determination of copper by this method are given below.

Experimental

Labelled material   As Experiment 24.1
Recommended activity 2 μCi

1. Prepare 100 cm^3 each of 0·01M solutions of nickel chloride and copper sulphate. These solutions will be used to prepare standard precipitates of labelled cupric phosphate (Cu$_3$(PO$_4$)$_2$), standard chromatograms and mixtures for separation and estimation of copper. The solutions contain 5·87 μg/0·01 cm^3 of Ni^{++} and 6·35 μg/0·01 cm^3 of Cu^{++}, respectively.

2. Prepare approximately 2 cm^3 of a labelled disodium hydrogen phosphate solution by adding 2 μCi of carrier-free phosphorus-32-labelled phosphate to 2 cm^3 of a 0·005M disodium hydrogen phosphate solution in a small beaker. About 0·05 cm^3 of this solution will precipitate up to 24 μg of copper.

3. Use an Agla syringe to dispense various volumes of the nickel and copper solutions onto paper chromatograms. Chromatographic paper (Whatman No. 1) may be used in the "Chinese lantern" form (25 cm × 25 cm) or as strips 2·5 cm × 25 cm. Prepare standard chromatograms by dispensing 0·01 cm^3 of each solution onto the origin of chromatogram paper. Prepare also chromatograms containing, say, 5, 10, 15 μg of copper together with about 10–20 μg of nickel. If successive additions of solution to the chromatogram are made, the solution must be air dried between additions and not

Experiment 26.3

more than 0·01 cm³ should be added at a time. Calibrated glass capillaries may be used instead of an Agla syringe.

4. Place the prepared chromatograms in a chromatography tank and develop them with an acetone/hydrochloric-acid/water solvent (86:6:8) using the ascending method. The solvent front should be allowed to rise about 10 cm (approximately one hour).

5. Remove the chromatograms from the tank and air dry them, marking the solvent front with a pencil. Place them in a tank with ammonia vapour for at least 30 min to neutralize the hydrochloric acid. Remove and again dry the chromatograms.

6. Locate the positions of Cu^{++} and Ni^{++} on the standard chromatograms with 0·1% rubeanic acid in ethanol and determine the R_f values. Copper forms a grey-green complex and nickel a grey-blue complex. Using the R_f values, locate the positions of the Cu^{++} ions on the other chromatograms. Cut out these regions with a 22 mm cork borer.

7. Cut several 22 mm filter paper discs and dispense amounts of the standard copper solution onto them containing approximately 5, 10, 15, 20 μg of Cu^{++}. Dry the discs as before.

8. Precipitate phosphorus-32-labelled cupric phosphate on the filter paper discs containing the standard and separated copper as follows. Place a 0·05 cm³ drop of the labelled phosphate solution on the centre of each disc, allow them to stand for about 30 s to ensure complete precipitation and then wash each disc in a small beaker containing distilled water for a further 30 s. Repeat this washing process twice to ensure complete removal of excess labelled phosphate and then dry the discs under an infra-red lamp.

9. Determine the counting rate for the labelled cupric phosphate precipitate on each disc with an end-window Geiger–Müller counter. Plot counting rate against copper content for the standards and use this graph to determine the copper contents on the test chromatograms.

Further work

Nickel may also be determined radiometrically by the same method.

Experiment 27

Photosynthesis

The green plant differs from most organisms in that it has the ability to reduce atmospheric carbon dioxide into complex molecules such as sugars. This is achieved by a photocatalytic reaction. Light supplies the energy required and chlorophyll pigments act as the catalyst. When light of the appropriate wavelengths (450 and 650 nm) is absorbed, the catalyst is activated and carbon dioxide is reduced to complex organic molecules and gaseous oxygen:

$$CO_2 + 2H_2O \xrightarrow[\text{+ light}]{\text{chlorophyll}} (CH_2O) + H_2O + O_2$$

According to van Niels (1935, 1945), the photochemical reaction is initially the splitting of two water molecules into (OH) and (H). These are not necessarily free radicals or atoms but are assigned the general terms oxidizing and reducing entities. The (OH) entities interact, producing oxygen and two further (H) entities:

$$(OH) + (OH) \rightarrow O_2 + 2(H)$$

Carrier molecules or hydrogen acceptors such as nicotinamide–adenine dinucleotide phosphate (NADP) become reduced by the (H) entities:

$$2NADP + 4(H) \rightarrow 2NADPH_2$$

and the $NADPH_2$ in turn reduces the carbon dioxide:

$$CO_2 + 2NADPH_2 \longrightarrow \quad \text{ATP (adenosine triphosphate)}$$
$$H_2O + (CH_2O) + 2NADP \longleftarrow \quad \text{ADP (adenosine diphosphate)}$$

This reaction requires energy for completion, and it is the molecule ATP with its energy-rich phosphate bonds that supplies it. From the basic sugar molecules, further complex materials are built up.

Carbon dioxide enters through the stomata of the leaves and is dissolved in the cell water to produce carbonic acid. This is partially neutralized by cations to bicarbonate ions (HCO_3^-). It is these that act as the reservoir of carbon dioxide for photosynthesis.

Experiment 27.1

In the following experiments, the path of carbon in photosynthesis is traced using the isotope carbon-14. Its half-life is 5760 years, and it decays with the emission of a weak β-particle. The use of this radioisotope to label carbon dioxide or bicarbonate ions enables an investigation to be made of its fixation into plants under various conditions. By chemical extraction of the plants and paper chromatography of the extracts, the fate of the labelled carbon atom may be followed. The selective labelling of particular regions of whole plants enables the translocation of the labelled sugars to be followed either with a portable Geiger–Müller counter or by the process of autoradiography. The latter is a technique whereby a radioactive specimen makes a radiograph of itself by its radiation interacting with a photographic emulsion.

References

1   VAN NIELS, C. B.   *Cold Spring Harbour Symposia Quant. Biol.*, **3**, pp. 138–150 (1935)
2   VAN NIELS, C. B.   *Advances in Enzymol*, **1**, pp. 268–328 (1945)

EXPERIMENT 27.1 FIXATION OF CARBON-14-LABELLED CARBON DIOXIDE BY PLANTS

Experimental

Labelled material   Solution of sodium carbonate containing 20 μCi/cm³ of carbon-14-labelled sodium carbonate and 20 mg/cm³ of sodium carbonate as carrier.
Carbon-14-labelled sodium carbonate (Radiochemical Centre catalogue number CFA 2)
Recommended activities 5–20 μCi
Plant material   Bean and tobacco leaves; variegated leaves such as *Geranium, Tradescantia, Coleus*, etc.

1. This experiment deals with a labelled gas, and all work should ideally be conducted in a fume cupboard, but failing this a well-ventilated room will suffice. The apparatus used is shown in Fig. E.27.1. Labelled carbon dioxide ($^{14}CO_2$) is generated in the bell-jar by running any dilute acid (approximately 1 cm³) from the thistle funnel onto 0·25–1 cm³ of the labelled sodium carbonate solution. It is necessary to ensure that the bell-jar and base-plate are well sealed with either Vaseline or Plasticine before carrying out the experiment.

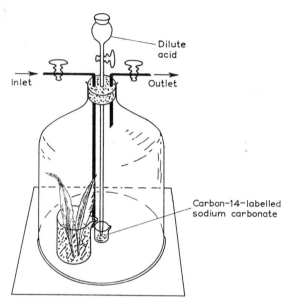

Fig. E.27.1 Apparatus for labelling plants with carbon-14-labelled carbon dioxide

If a fume cupboard is not available for the experiment, a partial vacuum should be applied to the chamber with the aid of a water or aquarium pump. In the event of a leak in the system, labelled gas would not then escape into the atmosphere. It is advisable not to overload the chamber with plant material, for this will result in very low activities in individual samples, due to possible shading and the distribution of the available activity over a large leaf area.

2. After generating the labelled carbon dioxide illuminate the belljar with a bright lamp for 2–3 h.

3. At the end of the photosynthesis period, pump out the excess carbon dioxide into 2M sodium hydroxide solution, usingthe in let and outlet tubes shown in the diagram. This is necessary when activities in excess of 1 μCi are used so that any radioactive waste may be disposed of as liquid waste. The Department of Education and Science regulations do not permit gaseous radioactive waste to be released into the atmosphere at a rate of more than 1 μCi/day.

4. Remove the plants with gloved hands or forceps and autoradiograph selected parts of the specimens as described in Experiment 27.2.

5. The presence of activity in leaf specimens may also be detected with a Geiger–Müller counter. Remove small discs from selected parts of the leaf with a cork borer, attach them to aluminium

Experiment 27.2 227

planchettes, dry and count under an end-window Geiger–Müller counter. Compare the count rates from samples kept in the light and shaded regions and from areas of leaves with and without pigments.

EXPERIMENT 27.2 THE TECHNIQUE OF MACROSCOPIC AUTORADIOGRAPHY USING PLANT MATERIAL

Ionizing radiations act upon photographic emulsion in a similar manner to light. If an X-ray film is exposed to an object containing radioactive material, then upon development of the film a photographic image providing visualization of the location of the radioactivity in the sample is obtained as shown in Plate 14. The process is termed *autoradiography*—literally, the sample makes a radiograph of itself. The smaller the distance between the specimen and the film, the greater is the resolving power of the autoradiograph. To obtain a satisfactory image, exposure times are considerably longer than those associated with conventional photography. Instead of seconds, as in photography, one is dealing with hours, days and perhaps months of exposure. The greatest problem is in fact judging the length of exposure time required to obtain a satisfactory result. Times are usually not critical, and Table 29 gives an idea of the

Table 29 Estimated exposure times for plant material labelled with carbon-14 for autoradiography with Kodak Industrex type D X-ray film

Counts per minute close to leaf surface*	Exposure time
500	3–4 d
2 000	24 h
5 000–10 000	10 h
20 000	2 h

* Measured with a Mullard MX 168 end-window Geiger–Müller counter (window thickness, 4–5 mg/cm^2)

general relationship between activity of sample and exposure times. Because of the limited penetration of carbon-14 β-particles, specimens containing the isotope must be placed directly on the photographic film.

The technique of setting up an autoradiograph is basically very simple; it is only necessary to place the sample and the film in close contact. There are, however, two situations causing artefacts (film images resulting from processes other than radiation images) which must be borne in mind. These are chemical artefacts produced as a result of cell exudates interacting with the emulsion, and pressure artefacts caused by raised positions of the sample (stalks, etc.)

being pressed into the emulsion, causing dislocation of the silver grains. These artefacts can look exactly like the true radiation-produced images.

Experimental

1. Arrange carbon-14-labelled plants prepared as in Experiment 27.1, spread out on filter paper or paper tissues. Sandwich them between metal sheets and dry in an oven at between 80° and 100°C. A weight may be necessary to keep the specimen perfectly flat. To prevent any chemical artefact due to interaction of the cell sap with the photographic emulsion, the plants must be perfectly dry. They will be brittle, however, and should be handled with care.

2. Using a Geiger–Müller counter, determine the count rate over various portions of the leaf, and from the table, estimate the exposure time.

3. Prepare a autoradiograph of the leaf, using Kodak Industrex type D envelope-wrapped X-ray film. In the dark-room, using a red safe-light (Wratten filter No. 1), remove the X-ray film from its packet and place the specimen directly onto the emulsion (both sides of the film are coated with emulsion). Do not apply heavy weights to the film in an attempt to get better apposition and thus better resolution, or pressure artefacts may be produced. Carefully replace the film and plant specimen in its light-proof wrapping and leave it for the appropriate exposure time.

4. At the end of the exposure time, process the film as follows:

   (i) Develop for 5 min in Kodak type D19(b) developer. To prepare the developer, dissolve in the order given

Metol	2·2 g
Sodium sulphite (crystalline)	144·0 g
Hydroquinone	8·8 g
Sodium carbonate (crystalline)	130·0 g
Potassium bromide	4·0 g
Distilled water	1000 cm^3

   (ii) Wash for 1 min in running water.
   (iii) Fix for 10 min in fixing solution (25–30% w/v sodium thiosulphate solution)
   (iv) Wash for 1 h in running water.
   (v) Dry.

EXPERIMENT 27.3 FIXATION OF INORGANIC CARBON INTO SIMPLE SUGARS BY PLANTS

The materials used in this experiment are those used in Experiment 27.1. Young broad bean or sunflower leaves, previously destarched, make suitable plant material.

Experimental

1. Label the plant, as previously described, by photosynthesis in 20 μCi of carbon-14-labelled carbon dioxide for 2 h. After this period check that activity is present in the leaves, using a Geiger–Müller counter.

2. Using a pestle and mortar, crush the leaves in 80% v/v ethanol/water, transfer the solution and crushed leaves to a flask and boil on a water bath for 10 min. Filter the sample, wash the residues with a little water and combine the filtrate and washings.

3. The sugars present in the leaf extract will be separated by paper chromatography. The presence of salts in the leaf extract often causes displacement and tailing of the spots on the chromatogram, and it is therefore necessary to remove these salts before preparing the chromatogram. This may be effected by passing the extract through an ion-exchange column. Since sugars are not ionized in solution, they are not retained on an ion-exchange resin whereas ionic species are retained.

Prepare two separate ion-exchange columns, one a cation (H^+) exchanger and the other an anion (OH^-) exchanger, by placing a glass-wool plug at the bottom of a 25 cm^3 burette, closing the tap and half-filling the burette with distilled water. Make a suspension of the resin with distilled water and add it to the column to make a resin bed of about 10 cm height. This method of preparation ensures that the resin bed is free from air bubbles. *Never allow the column to run dry*, or air bubbles and channels will form in the resin and it will have to be re-packed. The cation exchange resin may be purchased in the H^+ form, but the anion exchange resin is generally supplied in the Cl^- form. Before use this must be converted to the OH^- form by passing 50 cm^3 of M sodium hydroxide through the column, followed by 50 cm^3 distilled water.

4. Pour the extract onto the top of the cation exchange column, the outlet of which is connected to the top of the anion exchange column. Run the extract through both columns, followed by 20 cm^3 of distilled water. Collect the effluent from the columns.

5. Evaporate the column effluent to about 1 cm^3 and spot it onto Whatman No. 1 chromatography paper about 4 cm from the end of a strip 5 cm wide by approximately 40 cm long. Ensure that the material is confined to an area no greater than 0·5 cm in diameter and load the spot so that an activity of several hundred counts per minute is recorded by a Geiger–Müller counter placed above it. Spot aliquots of 1% w/v solutions of glucose, fructose and sucrose onto similar chromatograph strips. Develop by ascending chromatography using an isopropanol:pyridine:water (7:7:2) solvent, or by descending chromatography using n-butanol:ethanol:water (4:1·1:1·9) as solvent.

6. After developing and drying the chromatograms, mark the solvent fronts and stain the standards only with the following reagent:

m-phenylene-diamine	0·5 g
Stannous chloride	1·2 g
Acetic acid	20 cm³
Ethanol	80 cm³

On heating at about 80°C a brown colour develops at the sugar positions and enables the R_f values to be determined:

$$R_f = \frac{\text{Distance travelled by sample}}{\text{Distance travelled by solvent}}$$

7. Determine the R_f values of the sugars present in the labelled plant extract using either autoradiography or the chromatogram scanner shown in Plate 3. By the use of a collimator this apparatus enables the distribution of activity along chromatograms to be measured. Identify the sugars formed by comparison of the R_f values at which activity maxima occur with the R_f values of the standards.

8. The relative amounts of the sugars formed may be estimated from the relative activities of the chromatogram spots or from the relative blackness of their autoradiographs.

Note. Sugars are not easy to separate on a one-dimensional system. If time is available two-dimensional chromatography is far more satisfactory. Spot the concentrated sugar extract onto a 20 cm × 20 cm square of Whatman No. 1 chromatography paper, near the bottom left-hand corner. Similarly prepare standards of glucose, fructose and sucrose on separate squares of paper. Develop these chromatograms by ascending chromatography using a phenol-water solution. (28 g pure phenol in 12 cm³ water). Remove the chromatograms from the tank, dry them and then turn them through 90° so that the left-hand edge of the chromatogram is downwards. Develop them again by ascending chromatography using an *n*-butanol-acetic acid-water solvent. Prepare this solvent by mixing these reagents in the ratio 25:6:25 and use the upper layer formed.

EXPERIMENT 27.4 TRANSLOCATION OF LABELLED MATERIALS IN PLANTS

Some of the synthesized material in leaves is usually transferred from the site of production to the actively growing parts of the plant. The rate of this translocation can be conveniently measured using autoradiography. Quantitative evaluation is rather difficult, but three techniques for demonstrating the effect qualitatively are described below.

Experimental

1. A useful method of demonstrating translocation uses variegated leaves labelled by photosynthesis in the apparatus of Experiment 27.1. After labelling, leave the plant or leaf alive for approximately 24 hours, then dry and autoradiograph it. Compare this autoradiograph with that of a similar specimen killed immediately after labelling.

2(i) Using potted tomato or bean plants that have numerous side branches, enclose one branch or leaf in an atmosphere of carbon-14-labelled carbon dioxide using the apparatus in Fig. E.27.2.

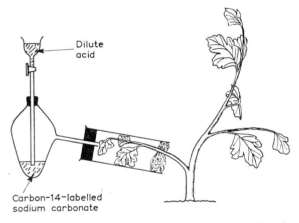

Fig E.27.2 **Method for selectively labelling side branches of a plant with carbon-14 by photosynthesis**

(ii) Place 5–10 μCi of carbon-14-labelled sodium carbonate in the flask, add some dilute acid and pass the labelled carbon dioxide generated into a chamber plugged with cotton-wool. Illuminate the apparatus for approximately 1 hour and then leave the plant in the fume cupboard for 1–7 days. Using a portable Geiger–Müller tube, trace the movement of activity in the plant over the period of growth. The final positions and relative activities of labelled areas can be estimated using autoradiography.

3. A localized region of the plant can be labelled by the technique of "spot" labelling using the glass J-tube shown in Fig. E.27.3.

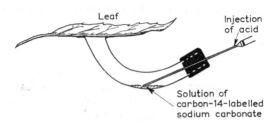

Fig E.27.3　J-tube method for spot labelling leaves by photosynthesis

(i) Grease the flattened section of the J-tube and apply it to a leaf of a pot-grown bean plant. Ensure there is good seal between the leaf and the J-tube.
(ii) Place about 2 μCi of carbon-14-labelled sodium carbonate solution in the tube and generate labelled carbon dioxide by injecting acid through a self-sealing serum stopper. Illuminate the plant for about 1 hour to allow photosynthesis to take place. Autoradiograph the plant after a further 1–7 days growth.

EXPERIMENT 27.5　THE INCORPORATION OF INORGANIC CARBON INTO ALGAE

Using single-celled algae, the fate of supplied inorganic carbon may be investigated. Labelled sodium bicarbonate is added to a culture of *Chlorella*, and after a period of synthesis the cells are extracted and the labelled organic molecules are identified by paper chromatography. Green algae are rich in proteins, and it is the constituent of proteins—amino acids—that are the main result of photosynthesis in these cells.

Experimental

Labelled material　　Sodium bicarbonate-C14 solution, 20 μCi/cm^3. (Radiochemical Centre catalogue number CFA 3)
Recommended activities　10 μCi
Plant material　　Culture of *Chlorella vulgaris* (approximately 7 days old) in 0·2 g of potassium nitrate, KNO_3, 0·02 g of potassium hydrogen phosphate, K_2HPO_4, and 0·02 g of magnesium sulphate, $MgSO_4 \cdot 7H_2O$, per 1000 cm^3.

1. Remove approximately 30 cm^3 of a *Chlorella* culture and keep in the dark for a few hours. Add 10 μCi of carbon-14-labelled sodium

Experiment 27.5

bicarbonate to the solution and illuminate it for approximately 20 min. A control kept in a darkened flask may also be prepared.

2. Filter and thoroughly wash the cells. Extract the cells with hot 80% ethanol for approximately 10 min. Filter and evaporate the filtrate to a small volume.

3. Load as much as possible of the extract on to the Whatman No. 1 chromatography paper, using a small loop of wire, and dry the paper after each application. This ensures a limited spread of the samples—an important requirement for good chromatographic separation.

4. Spot standards onto the paper (aspartic acid, alanine, glycine and valine) and develop by ascending chromatography using an ethanol:ammonia:water solvent (8:1:1). Ensure that only a small amount of the standard amino acids are applied to the paper. The colour test for them at the end of the experiment is extremely sensitive. The chromatogram takes approximately 3–4 hours to run.

5. Dry the chromatography paper and spray with ninhydrin. A blue colour will develop on drying at about 110°C. This is a characteristic colour reaction of amino acids. Cut off the extract strip from the main sheet of paper and determine the radioactivity distribution along it with a chromatogram scanner (Plate 3). Compare the R_f values of the standards with those of activity peaks of the extract and so identify the amino acids present in the extract.

In most cases the amino acids present in *Chlorella* extract will stain blue, and activity can be seen to be associated with these areas. An alternative method of estimating the position of the labelled amino acids is to autoradiograph the separated extract. Because of the fairly low count rates, exposure times may be of the order of 1–2 weeks.

Note. A simpler experiment may be conducted with this system, in that, after labelling with carbon-14, instead of separating the amino acids, the control and experimental samples are simply filtered and washed on a demountable filter (Fig. 6.3). They are then dried and counted under an end-window Geiger–Müller tube, and the activities compared. The influence of light of different wavelengths on the fixation of carbon-14 may also be investigated using this experiment.

Experiment 28

Respiration

Respiration is a reversal of photosynthesis in that it is the means by which complex food materials, such as glucose and fructose, are broken down into simpler substances. The resultant release of energy is utilized by the plant for growth and all activities associated with it.

In the following series of experiments carbon-14-labelled organic molecules are broken down into an easily measured waste product, namely carbon-14-labelled carbon dioxide. Labelled gas may be estimated directly using a Geiger–Müller tube with a gas-tight chamber surrounding the thin end-window as shown in Fig. E.28.1. This apparatus is used in Experiment 28.1, but it must be pointed out that it is purely a monitoring device—very little quantitative work can be done with it. To make direct measurements of the carbon-14 evolved, the carbon dioxide must first be absorbed in sodium hydroxide and then precipitated as barium carbonate. The precipitate is then counted under an end-window Geiger–Müller counter. Because of the self-absorption problems associated with the low-energy β-particles from carbon-14, a correction has to be made (see Experiment 16).

In part of Experiment 28.1 the output of labelled carbon dioxide is measured from a mouse fed on carbon-14-labelled glucose. The main problem in such an experiment is how to get the labelled material into the animal without infringing the Cruelty to Animals Act, 1876. The following alternative procedures are available:

(i) To inject the glucose intraperitoneally
(ii) To force feed the animal through a plastic tube
(iii) To put glucose into its drinking water
(iv) To spot labelled glucose onto a small food pellet

A Home Office licence is required for methods (i) and (ii). The Secretary of State for the Home Department is of the opinion, however, that the procedure whereby an animal is fasted for 24 hours and then fed radioactive glucose is not an experiment within the meaning of the Cruelty to Animals Act, since the result would be predictable, and it appears unlikely that the procedure in itself would be liable to cause pain to the animal.

Experiment 28.1

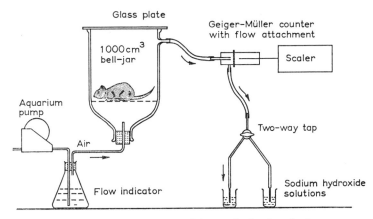

Fig E.28.1 Apparatus for studying respiration in animals

EXPERIMENT 28.1 RESPIRATION IN PLANTS AND ANIMALS

Experimental: Plant Respiration

In this experiment, a plant, previously labelled with carbon-14 dioxide as described in Experiment 27 can be conveniently used as a source of carbon-14-labelled food stores. As the experiment can last over several days, a potted plant is essential.

1. Confine a carbon-14-labelled plant to a small bell-jar together with a beaker containing approximately 5 cm^3 of freshly prepared 2M sodium hydroxide. Ensure effective sealing of the unit, and place it in a dark cupboard.

2. Remove the beaker of sodium hydroxide regularly and replace it with a fresh supply. Precipitate completely the carbonate formed, using a mixture of M barium chloride and M ammonium chloride. Filter or preferably centrifuge, wash and suspend the precipitate in acetone. Transfer to a weighed planchette, dry and reweigh. Count the source with an end-window counter.

3. Correct the count rates for self-absorption from a calibration curve prepared as described in Experiment 16. Plot the specific activity of the barium carbonate (as counts per minute per milligram) against time.

Counting may also be made over selected leaves, and the loss of activity from them can be correlated reasonably well with the uptake of carbon dioxide by the sodium hydroxide. Two processes are being measured in this method, however, respiration and translocation.

The major output of labelled carbon dioxide occurs within the first 24 hours. After that time there is only a very slow output.

Experimental: Animal respiration

Labelled material — Carbon-14-labelled glucose absorbed on filter paper. Recovery of the glucose is carried out by the method described in Note 3 below. 50 μCi D-Glucose-C14 (U) (Radiochemical Centre catalogue number CFB 35).

Recommended activity — 10 μCi carbon-14-labelled glucose plus carrier glucose.

Animal — Mouse if glucose is to be injected. (A Home Office licence is required before animals may be injected.) Rat or guinea pig if glucose is added to drinking water.

1. Set up the apparatus as shown in Fig. E.28.1. The scaler and Geiger–Müller counter are useful but not absolutely necessary. They are used as a monitoring device to register the start of radioactive gas evolution. The exhaled gas is trapped in approximately 5 cm^3 of freshly prepared 2M sodium hydroxide placed in the bubbler as shown.

2. Feed the animal or inject it with a solution containing approximately 10 μCi of carbon-14-labelled glucose and 1 g of carrier glucose. Then place it in the respiration chamber, ensuring gas-tight sealing.

3. Remove the sodium hydroxide solution at regular intervals and replace it with a fresh aliquot. Precipitate barium carbonate and recover the precipitate on a weighed planchette. Dry and weigh the prepared source and then determine its activity with an end-window counter.

4. Correct the count rates for self-absorption from a calibration curve prepared as described in Experiment 16. Plot the specific activity of the barium carbonate (as counts per minute per milligram) against time.

Most of the labelled carbon dioxide will be evolved within the first 2 hours. Any urine that may have collected in the bottom of the chamber can be tested for activity by evaporation to dryness and counting under an end-window counter.

Notes

1. In these experiments, specific activity is always plotted against time because the whole of the precipitate is seldom used. If all the precipitate were mounted on the planchette, the total activity could be plotted against time. This, however, is time consuming.

Experiment 28.1

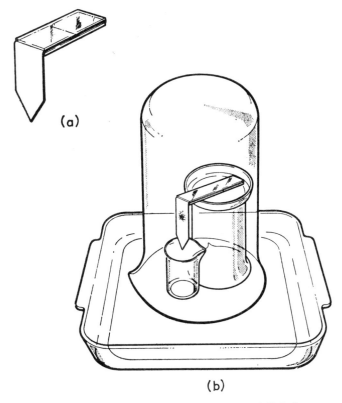

Fig E.28.2 Method for the elution of carbon-14-labelled glucose from chromatograph paper

(Reproduced with permission of the Radiochemical Centre Ltd.)

2. Conduct the experiment in a well-ventilated room or preferably in a fume cupboard. At the end of the experiment, confine the animal to a cage outside for 2–3 days, so that all the carbon-14 activity is excreted.

3. Recovery of labelled glucose spread on filter paper: open the ampoule by making a scratch mark with a glass-knife close to the top of the tube and applying to the mark a white-hot glass rod. The scratch mark will fracture the glass and make opening easy. Remove the filter paper with gloved hands, unroll it and position it between two microscope slides (Fig. E.28.2(*a*)). Set the end of the slide in a few cubic centimetres of distilled water (Fig. E.28.2(*b*)) and elute off the glucose. Within about an hour almost all the activity will have been removed. Ensure that the whole unit is covered with a bell-jar or beaker to minimize evaporation.

EXPERIMENT 28.2 THE INFLUENCE OF TEMPERATURE ON THE RESPIRATION OF BACTERIA

Bacteria grown in a medium supplemented with labelled substrates may selectively break down the substrates to easily determined end products. Substrates that may be used are carbon-14-labelled glucose, malic acid and sodium formate. Their eventual breakdown products are carbon-14-labelled carbon dioxide plus a variety of mainly unlabelled material. By measuring the rate of carbon dioxide evolution, the metabolic behaviour of the cells may be assessed. The metabolic activity of the organisms is usually related to their rate of replication. All the substrates chosen are involved in carbohydrate metabolism, and in the case of glucose and malic acid, all the carbon atoms of the molecule are labelled.

Experimental

Labelled material Sodium formate—C-14 CFA 11 100 μCi
Glucose—C-14 CFB 35 50 μCi
Malic acid—C-14 CFB 42 50 μCi

Recommended activity Approximately 2 μCi/cm^3 of broth or basal medium.

Material Culture of coliform bacteria (*Escherichia coli*) McConkey broth (Oxoid) or basal medium containing per 1000 cm^3:
Dipotassium hydrogen phosphate 7 g
Potassium dihydrogen phosphate 3 g
Sodium citrate 0·5 g
Magnesium sulphate 0·1 g
Ammonium sulphate 1 g
Add 5 cm^3 of 20% w/v glucose solution per 100 cm^3 of the above medium, the glucose and medium being sterilized separately

The unit for culturing the bacteria and trapping the labelled carbon dioxide is shown in Fig. E.28.3.

By inoculating the labelled medium with a culture of coliform bacteria and incubating the unit for a period of time, the labelled carbon dioxide evolved will be absorbed by the saturated barium hydroxide in the upper planchette. After the incubation period the hydroxide is evaporated to dryness and counted under an end-window Geiger–Müller counter.

Experiment 28.2

1. Pipette three 0·5 cm³ samples of labelled medium into aluminium planchettes. Inoculate each with 0·005 cm³ of an overnight culture of *E. coli*.
2. Place these inoculated samples in the apparatus shown in Fig. E.28.3. The upper nickel-plated planchette should contain 0·5 cm³ of saturated barium hydroxide solution. Screw the metal

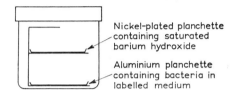

Fig.. E.28.3 Culture unit for investigating the metabolism of carbon-14-labelled substrates by bacteria

cap down tightly and incubate the units at room temperature, 30° and 40°C.

3. At intervals of about 30 min, remove the upper planchette and replace it with another containing a fresh aliquot of barium hydroxide. Evaporate the hydroxide solutions that have been removed and count the dried samples under an end-window Geiger–Müller counter.
4. Plot $\log_{10}$ (accumulated count rate) against time and correlate the variation of metabolic rate with temperature.

Notes

1. The volume of the inoculum is 0·005 cm³, and dispensing this requires an accurate pipette or syringe. The advantage of putting exactly the same volume of bacteria in each culture unit is that the metabolic curves start at the same level of activity. Should such a pipette not be available, then a wire loopful of inoculum may be dispensed into each sample of labelled medium. With such a procedure the reproducibility of the numbers of bacteria inoculated cannot be guaranteed, but the rate of metabolic activity is generally dependent on temperature and not initial numbers of cells.
2. At the end of the experiment, place the growing cells in disinfectant.
3. The effects of various metabolic inhibitors such as cyanide and azide may be investigated with this apparatus.
4. As we are interested only in the rate at which the accumulated activity increases, we need to know only the relative counting rates of the different sources. In this case a self-absorption correction need not be made to each count, for the weight of barium carbonate precipitate will be the same for each sample (see Experiment 16).

Experiment 29

Mineral nutrition and ion movement across membranes

In addition to the elements oxygen, hydrogen and carbon, which are basic requirements for growth, there are a large number of mineral elements that occupy key positions in many biochemical systems and are therefore essential for both growth and development of the whole organism. The necessity for the various mineral elements in plants has been thoroughly investigated by the use of water culture solutions lacking particular minerals. The experiments show the effect on a growing plant of the lack of these elements. To gain further understanding of the mechanism involved in ion uptake, radioactive tracers have been extensively used. With these the kinetics of ion movement in cells is now becoming clear. Fig. E.29.1 is a general summary of the mineral uptake mechanism. The two principal membranes of plant cells are the outer cytoplasmic membrane, or

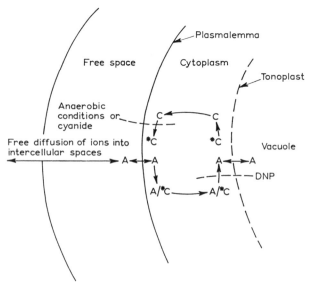

Fig E.29.1 Diagrammatic representation of ion uptake by the root system of plants

plasmalemma, and the inner cytoplasmic membrane bordering on the central vacuole, the *tonoplast*. The vacuole is the main repository for accumulated ions, and from dilute solutions of salts, both cations and anions may be absorbed until the concentrations in the vacuole are in excess of the external concentrations.

It is now well documented that metabolic reactions are necessary to furnish energy for the accumulation of ions, for in general when normal metabolism is inhibited ion accumulation is interfered with. Lundegårdh considered that there was a strict connection between respiration and anion movement. However, 2:4-dinitrophenol (DNP) was found to increase the respiratory rate and inhibit salt absorption, a fact difficult to reconcile with Lundegårdh's hypothesis. It now appears more probable that oxidative phosphorylation has the necessary role in ion transport. Both respiration and phosphorylation are closely linked; respiration in fact supplies the energy for the production of adenosine triphosphate (ATP). DNP acts as a decoupling reagent, so that its addition to the system allows respiration to continue but stops the manufacture of ATP. Experiments on the effects of DNP on the absorption of rubidium by plant material support the conclusion that absorption requires the availability of ATP.

A model to account for ion absorption is included in Fig. E.29.1. An active carrier (*C) present in the cytoplasm combines with the ion (A) at the outer side of the membrane; the resulting carrier-ion complex (A/*C) crosses the cytoplasm (considered to be impermeable to the free ion) and through a chemical change in the carrier molecule (*C → C) is released into the inner space or vacuole. The ion is now not free to return to the external solution because of the impermeability of the membranes to the non-complexed ion and the lack of affinity of the changed carrier (C) for the ion. The carrier therefore operates in a cyclic fashion, and it is the carrier rather than the ion that undergoes a chemical change in the absorption process. From the diagram it can be seen that anaerobic conditions and respiratory poisons (cyanide) inhibit the carrier system, and DNP prevents the uncoupling of the carrier ion complex.

The following experiments are designed to show how simple investigation may be made on the uptake of phosphate and sulphate ions in whole plants and in root systems. Using a series of competitors and inhibitors, some information may be gained on the reactions necessary for ion uptake.

EXPERIMENT 29.1 THE UPTAKE AND TRANSLOCATION OF PHOSPHORUS-32-LABELLED PHOSPHATE IN PLANTS

Phosphorus-32-labelled phosphate incorporated in plant tissues is easily detected with an end-window Geiger–Müller counter. Translocation of the isotope can therefore be qualitatively measured very simply by searching for the labelled regions of a plant. However, since phosphorus-32 emits a very energetic β-particle, the counter has to be adequately shielded from the interference of both the stock nutrient solution and radioactive regions of the plant other than those under observation. In practice this is difficult to achieve, and a more effective way of following the fate of the phosphate at the end of uptake is to examine the plant by autoradiography.

The uptake of the labelled phosphate is effected by:

(i) Immersing the roots of the plant in a solution containing the tracer (Fig. E.29.2(a)).

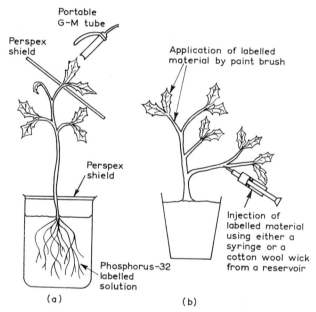

Fig E.29.2 Methods for labelling plants with phosphorus-32-labelled phosphate

(ii) Introducing the tracer half-way up a side branch of the plant with the aid of a hypodermic needle and syringe (Fig. E.29.2(b)).
(iii) Painting the tracer solution on either side of selected leaves and demonstrating foliar absorption and efficiency of translocation (Fig. E.29.2(b)).

All these techniques have been used extensively in comparisons of a variety of phosphate fertilizers.

Experiment 29.1 243

Experimental

Labelled material   Phosphorus-32-labelled phosphate (Radiochemical Centre catalogue number PBS 1).
Recommended activity 1–10 µCi
Plant material   Pot-grown tomato plants, 6–12 in high

1. *Labelling the Plant*

Method (i). Remove a tomato plant from its pot and wash its root system free of soil. Place the plant in a beaker containing about 150 cm³ of tap water to which about 10 µCi of phosphorus-32-labelled phosphate and a few milligrams of carrier phosphate have been added. Cover the solution and root system with lead foil or Perspex sheet by fitting it to the top of the beaker.

Method (ii). Inject about 1 µCi of carrier-free phosphorus-32-labelled phosphate in about 0·1 cm³ into the stem of a side branch on a pot-grown tomato plant. The plant does not need to be removed from the pot.

Method (iii). Prepare a solution containing about 10 µCi of phosphorus-32-labelled phosphate and a few milligrams of carrier phosphate in 0·5 cm³. Paint sufficient of this solution to contain 1–2 µCi onto selected leaves of a pot-grown tomato plant.

2. *Following translocation*

Over a period of about 48 hours, occasionally investigate the positions in the plants at which phosphorus-32 activity can be found, using an end-window Geiger–Müller counter as shown in Fig. E.29.2(a).

3. *Autoradiography*

The technique of autoradiography has already been described for carbon-14-labelled samples (Experiment 27.2). In the case of samples labelled with phosphorus-32 the much more energetic β-particle emitted makes it unnecessary to dry the plant material or to remove the X-ray film from its envelope packing. With an undried sample, however, a piece of polythene film must be used, as in Fig. E.29.3 to

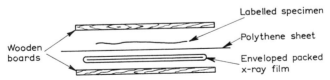

Fig E.29.3   Method for autoradiography of a phosphorus-32-labelled specimen

The wooden boards are held together by elastic bands to ensure close apposition between the X-ray film and specimen

prevent the production of chemical artefacts on the film. To ensure close apposition between the plant and the film, elastic bands may be used to hold the assembly together. It is inadvisable to use heavy weights for this purpose, as distortion of the emulsion by the raised parts of the specimen (stem, veins, etc.) will also cause artefacts.

Prepare selected parts of the plants for autoradiography as described above. The exposure time necessary for the activities used will be about 1 week. Develop the films as described in Experiment 27.2.

Further work

The uptake of sulphur-35-labelled sulphate and its translocation can be conducted in the same way. Because sulphur-35 emits a very weak β particle, the sample for autoradiography must be dried in the oven at 100°C and then placed next to the photographic emulsion.

Autoradiographs of higher resolution are obtained with isotopes which emit weak β-particles. This is because the low-energy β-particles are stopped in the emulsion within a short distance of their point of origin in the specimen. A high-energy β-particle, such as that from phosphorus-32, may sensitize the emulsion at comparatively great distances from its origin in the specimen so that the autoradiograph cannot localize the origin of the particles so precisely.

EXPERIMENT 29.2 DETERMINATION OF THE RATE OF UPTAKE OF PHOSPHATE AND SULPHATE IONS BY PLANTS

A general introduction to the method of tracing phosphorus-32-labelled phosphate uptake in plants is described in Experiment 29.1. The uptake was qualitatively measured using a portable end-window Geiger–Müller counter and autoradiography. The present experiment is a modification of this, in that direct measurement on the plant is unnecessary; instead, loss of activity from a nutrient solution is measured. This gives a better indication of the rate of uptake of ions by the plant over a period of time.

The experiment also serves to demonstrate a method of distinguishing the rate of uptake of two different ions in the presence of one another. The ions used are sulphur-35-labelled sulphate and phosphorus-32-labelled phosphate. As sulphur-35 emits a β-particle of much lower energy than phosphorus-32 it is possible in the process of counting to distinguish between both sources, by absorption of the β-particles from sulphur-35 in a suitable aluminium absorber. The experiment includes a study of the effects of aeration and de-aeration of the nutrient solutions on ion uptake.

Experiment 29.2

Experimental

Labelled material Phosphorus-32-labelled phosphate (Radiochemical Centre catalogue number PBS 1). Sulphur-35-labelled sulphate (Radiochemical Centre catalogue number SJS 1).

Recommended activity 8 µCi of phosphorus-32 and 12 µCi of sulphur-35.

Plant material Oats or barley seedlings approximately 14 days old, grown in a nutrient solution lacking both phosphate and sulphate ions. For each experiment, take two $2\frac{1}{2}$ in plastic pots, cut out the bottom to leave a small rim and replace it with plastic meshwork as shown in Fig. E.29.4. Plant 20 seeds in each container and

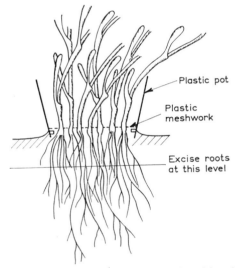

Fig E.29.4 Method for germinating and excising the roots of barley seedlings

bring them into close contact with the aerated nutrient solution (1·2 mM of potassium nitrate, 0·5 mM of calcium nitrate, 0·2 mM of magnesium chloride, and trace amounts of zinc, copper, iron, molybdenum, boron and manganese in 1000 cm³). Cover them with a polythene sheet. The cover should be removed when the green shoots are approximately 3 cm long. To ensure a steady state, renew the nutrient solutions frequently.

1. Transfer two 200 cm³ aliquots of nutrient solution to two 600 cm³ beakers. Add approximately 8 μCi of phosphorus-32-labelled phosphate and 12 μCi of sulphur-35-labelled sulphate (both carrier free) to both beakers and mix thoroughly. These are the experimental solutions.

2. Remove a 1 cm³ aliquot for activity determination and evaporate it to dryness on a planchette.

3. Transfer approximately 2 × 20 plants to fresh inactive nutrient solution in two beakers, aerate one using an aquarium aerator, and outgas the other with nitrogen. Continue this for approximately 30 min to achieve a steady state as regards oxygen tension in the roots.

4. Aerate and outgas the labelled experimental solutions for a few minutes and then immerse the prepared plants. Continue bubbling air and nitrogen to the solutions throughout the experiment.

5. Remove 1 cm³ aliquots of the solutions for counting at the following intervals after immersion of the plants: 2, 10, 20, 30, 60, 120 and 240 min, placing them on counting planchettes.

6. Remove the plants, blot them dry, and weigh them.

7. The planchettes, after drying under the infra-red lamps, are counted with an end-window Geiger–Müller counter, and then counted again with the same counter without change of geometry but with a 40 mg/cm² aluminium absorber placed between the sample and the counter window. This absorbs the sulphur-35 β-particles (see Experiment 7).

8. From the decrease in activity of the solutions, calculate the increase in activity per unit weight of plant material and plot it against time.

Note. The Department of Education and Science maximum permissible levels of activity of phosphorus-32 and sulphur-35 per experiment are respectively 16 and 48 μCi. If both isotopes are used together as in this experiment, the maximum permitted quantities of each are 8 and 24 μCi. The experiment must therefore be conducted as two separate experiments, one with aeration and one with de-aeration of the nutrient solution.

After the experiment the plant material and the remaining nutrient solution will have to be disposed of as radioactive waste. About half of the original activity will have been absorbed by the plant, and care must be taken not to exceed the level of 10 μCi of solid waste which may be disposed of under the Radioactive Substances Act, School Exemption Order, by those institutions working under the D.E.S. regulations. If more than one experiment has been done it may be necessary to carry out a wet oxidation of the plant material with concentrated nitric acid and then make the waste disposal as liquid waste after dilution with water. The remaining nutrient solution may be disposed of as liquid waste.

Experiment 29.3 247

References

FRIED, M., *et al.* "Effect of reduced oxygen tension on the uptake of inorganic ions by rice and barley", *Symposium on Isotopes and Radiation in Soil-Plant Nutrition Studies*, p. 233 (International Atomic Energy Agency, Vienna 1965) The above experiment is a modification of one described in *Laboratory Training Manual on the Use of Isotopes and Radiation in Soil-Plant Relations Research* (International Atomic Energy Agency, Vienna 1964) and is included here by permission of the publishers.

EXPERIMENT 29.3 INVESTIGATION OF THE EFFECTS OF INHIBITORS AND COMPETITORS ON THE UPTAKE OF IONS BY EXCISED ROOTS

This experiment is a development of the previous one in that the influence of 2:4 dinitrophenol and sodium selenate on the uptake of sulphate ions is investigated. Oats or barley plants grown in a nutrient medium deficient in sulphate have their roots excised and immersed in labelled solutions. The decrease in activity of the solutions is used to calculate the ion uptake per unit weight of root tissue. The roots are transferred to inactive solutions (tap-water or calcium sulphate solution) after the uptake experiment and the release of activity into this solution is determined. This permits the removal of the diffusible ions in the root system to be followed. The presence of activity retained in the roots by active transport is subsequently measured by wet oxidation of the roots and measurement of the activity in the resulting solution.

Experimental

Labelled material   As for Experiment 29.2
Recommended activity 20 μCi of sulphur-35-labelled sulphate or 10 μCi of phosphorus-32-labelled phosphate (see Note 3)
Plant material   As for Experiment 29.2

1. From 14-day-old seedlings grown as in the previous experiment, cut off the roots which have grown through the plastic mesh into the nutrient solution as in Fig. E.29.4. Blot the roots dry and weigh four samples each of about 1 g fresh weight. Immerse three of the root samples in distilled water and aerate them for about 1 hour. Kill the fourth sample by immersion in boiling water.
2. Place the root samples in labelled solutions as follows:
   Samples 1 and 2   50 cm³ nutrient solution containing 0·1 μCi/cm³ of sulphur-35-labelled sulphate.
   Sample 3   50 cm³ nutrient solution containing 0·1 μCi/cm³ of sulphur-35-labelled sulphate and 0·2 mg/cm³ of sodium selenate.

Sample 4   50 cm³ nutrient solution containing 0·1 μCi/cm³ of sulphur-35-labelled sulphate per cm³ and 10^{-4} M 2:4 dinitrophenol.

Use sample 2 for the killed roots.

3. Note the time at which the samples are placed in the labelled solutions and aerate the solutions throughout the experiment.

4. Transfer 0·25 cm³ aliquots of each labelled solution to counting planchettes at zero time (before immersion), and at 4, 8, 15, 25 and 35 min after immersion. Evaporate to dryness and count each under an end-window Geiger–Müller counter.

5. Remove the roots from the solutions after 35 min, blot dry, wash rapidly with distilled water from a wash bottle, and immerse them in tap-water or a calcium sulphate solution (20 mM/1000 cm³) for 1 hour.

6. Transfer the roots to separate dry beakers, add a few cubic centimetres of concentrated nitric acid and boil to oxidize the roots.

7. Evaporate the tap-water and oxidized root samples to dryness. To each, add 1 cm³ of water, transfer to planchettes, evaporate to dryness and count.

Notes and further work

1. Plants require minute amounts of sulphate ions and it is essential to be accurate in the timing and sampling of the aliquots.

2. 10^{-4} M potassium cyanide may be used in place of 2:4 dinitrophenol.

3. Conduct the same experiment for phosphate uptake, using the following labelled solutions:

Samples 1 and 2   50 cm³ nutrient solution containing 0·05 μCi/cm³ of phosphorus-32-labelled phosphate and 0·5 mg of carrier phosphate.

Sample 3   50 cm³ nutrient solution containing 0·05 μCi/cm³ of phosphorus-32-labelled phosphate, 0·5 mg of carrier phosphate and 10^{-4} M 2:4 dinitrophenol.

Sample 4   50 cm³ nutrient solution containing 0·05 μCi/cm³ of phosphorus-32 labelled phosphate, 0·5 mg of carrier phosphate and 10^{-5} M sodium hydroxide.

Experiment 30

The technique of microscopic autoradiography for studying the localization of radioisotopes in tissue

The basic concept of autoradiography together with the technique has been considered previously (Experiment 27.2). The particular experiments described dealt with macroscopic autoradiography. In these, an investigation was made of the general distribution of labelled material over a leaf. The technique of microscopic autoradiography, whilst in principle the same, deals with the localization of radioisotopes at the tissue and cellular level. A histological section, labelled with a radioisotope, is placed in close apposition with a sensitive photographic emulsion. After a suitable period of exposure the emulsion and specimen are processed together without being separated as for macroscopic work. It is, in fact, essential that no displacement takes place between the emulsion and specimen, because in such experiments where localization at the cellular level is being investigated a film movement of a few micrometres can seriously affect the results. The preparation of the histological section is a conventional technique and will not be discussed here. The setting up of this type of autoradiograph, however, requires further elaboration.

Resolution is important with this work, for when observing the completed autoradiograph under the microscope, the silver grains of the autoradiograph must be directly above the labeiled tissue or cells. To achieve the best resolution possible, the sensitive emulsion must be right next to the specimen. A small air gap between specimen and emulsion would seriously affect the resolution. There are two basic ways of efficiently applying the emulsion to the section. They are (i) to use a photographic emulsion in the liquid form and dip the slide into it, (ii) to use a thin layer of emulsion and wrap it round the prepared slide. It is the latter technique—known as the *stripping film technique*—that will be used in this experiment. The thin layer (5 μm) of emulsion used is supported on a glass plate with a 10 μm layer of gelatin sandwiched between the emulsion and the glass (Fig. E.30.1(*a*). The film is stripped off the glass support and floated on water. In this wet state a prepared slide can be submerged under the floating emulsion and lifted in such a way that the film wraps itself neatly round the slide.

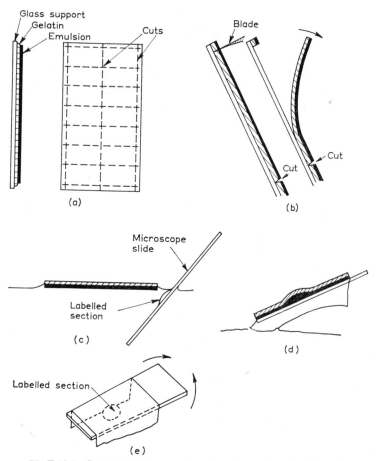

Fig E.30.1 Stages in the preparation of a microscopic autoradiograph

The experiment chosen to demonstrate this particular technique of autoradiography is an investigation of the uptake and distribution of iodine-131-labelled sodium iodide in the tissues of the tadpole. Plates 15(a) and 16(a) are photomicrographs of a thyroid section, and Plates 15(b) and 16(b) are autoradiographs of a similar section labelled with iodine-131. It may be noted that the autoradiographs are composed of small black silver grains and that their distribution in the tissue is related to the presence of iodine-131.

Experimental

Labelled material   Iodine-131-labelled sodium iodide in sodium thiosulphate solution (Radiochemical Centre catalogue No. IBS 1P).

Experiment 30

Recommended activity 1·5 μCi. This is the maximum that may be used per experiment.
Material Tadpoles of *Rana* or *Xenopus* at various stages of metamorphosis.

Preparation of Labelled Animals
1. Immerse the tadpole in 3 cm³ of water containing about 1 mg of inactive sodium iodide and about 1·5 μCi of iodine-131.
2. Leave overnight, wash several times to remove any labelled isotope adhering to the outside of the animal, kill in MS 222 and fix in Bouin's fixative (75 cm³ of a saturated aqueous solution of picric acid + 25 cm³ formalin + 5 cm³ glacial acetic acid). Keep in fixative for about 10 h.
3. Take up through the alcohols as follows:
   (a) Several changes 70% alcohol to remove all picric acid (15 hours).
   (b) 90% alcohol (3 h).
   (c) Absolute alcohol—2 changes (2 h).
   (d) Xylene—3 changes (3 h).
   (e) Paraffin wax (m.p. 49°C) at about 50°C—3 changes (3 h).
4. Cut sections about 6 μm thick and mount them on clean microscope slides that have previously been "subbed" in a gelatin solution (5 g of gelatin and 0·5 g of chrome alum in 1000 cm³ of water).
5. Bring the sections down through the following solutions:
   (a) Xylol (10 min).
   (b) Absolute alcohol (1 min).
   (c) Absolute alcohol (1 min).
   (d) 90% alcohol (1 min).
   (e) 70% alcohol (1 min).
   (f) 50% alcohol (1 min).
   (g) 30% alcohol (1 min).
   (h) Two changes of distilled water (30 min).

Preparation of the autoradiograph
Fig. E.30.1 illustrates the general principles of setting up the autoradiograph. The autoradiographic film used is Kodak AR10, which is universal for this type of autoradiography.

1. In the darkroom, using a red safelight (Wratten No. 1), make razor cuts through the emulsion as shown in Fig. E.30.1(a). Each oblong section between the cuts will be used to cover the sections of one slide.
2. With a thin blade, prize up the edge of an oblong section of the film and gently strip it off the glass as shown in Fig. E.30.1(b). The

relative humidity of the darkroom should be maintained at about 65%, because the act of stripping the film in a dry atmosphere can generate electrostatic charges which produce light flashes causing fogging of the film. A water bath maintained at 37°C will obviate this effect.

3. Float the stripped piece of film onto the surface of fresh distilled water and leave for about 3 min to expand. *Ensure that the emulsion is facing downwards.* Gently lower the microscope slide under the floating emulsion (Fig. E.30.1(c)), and lift it so that the emulsion wraps itself evenly over the section, as at (d). The emulsion layer will now be directly against the labelled specimens. Twist the slide so that the loose flaps of film stick to the underside of the slide, as at (e).

Dry the slide with cool air and leave it for exposure in a light-proof box containing silica gel. Store in a cool place. The exposure time for the activities used in this experiment is about 7–10 days.

Film processing
 (i) Develop the film for 5 min in filtered D19b developer at 20°C.
 (ii) Rinse in fresh distilled water for 15 s.
 (iii) Fix in 20–30% sodium thiosulphate for 10 min. Ensure the film is quite clear.
 (iv) Wash in distilled water for 30 min.

Staining the section

	minutes
(i) Stain in Ehrlich's haematoxylin.	30
(ii) Differentiate in 0·2% v/v concentrated hydrochloric acid in water	1–1½
(iii) "Blue" in running tap water	30
(iv) Wash in distilled water	5
(v) Stain in aqueous Eosin solution	2–3
(vi) Differentiate in distilled water	5

 (vii) Dry slides, stripping off excess film under the slide.
 (viii) Mount under cover slip with DPX mountant.
 (ix) Examine, thyroid, kidneys, gut, liver, etc. for localization of iodine.

Notes
 1. The temperatures quoted in the experiment must be strictly adhered to. Development and staining of the film and sectioning at too high a temperature may result in displacement of the film in relation to the specimen.
 2. Cleanliness is essential in such microscopy experiments.

Experiment 30

3. The autoradiographs must be exposed in the dry state. If exposed under moist conditions, latent images on the emulsion may be lost.
4. The thyroid becomes active when the hind legs of the animals are 3–4 cm long—prior to that little activity will be found there.
5. Formula for D19b: see page 228.

Appendix 1

Glossary of nuclear terms

Absorber. Any material that absorbs or "stops" radiation. Strong neutron absorbers like boron, and cadmium are used in reactor control rods. Lead, concrete and steel attenuate γ-rays and neutrons. Aluminium, Perspex and glass effectively absorb β-particles. Paper stops α-particles.

Activation. The process of making a material radioactive by bombardment with neutrons, protons or other particles.

Activity. The rate of disintegration of a radioactive material. The unit is the curie (Ci), which equals $3 \cdot 7 \times 10^{10}$ nuclear disintegrations per second.

Alpha particle (α). A positively charged particle emitted by certain radioactive materials. It is made up of two neutrons and two protons; hence it is identical with the nucleus of a helium atom. It is the least penetrating of the three common forms of radiation (alpha, beta, gamma), being stopped by a sheet of paper. A nucleus which loses an α-particle is transformed to an isotope of the element two places lower in the periodic table.

Annihilation. The process whereby an electron and a positron disappear or annihilate, producing two γ-rays, each with the energy equivalent of the rest mass of an electron, i.e. 0·51 MeV. The γ-rays are emitted in opposite directions.

Antimatter. Matter in which the ordinary nuclear particles (neutrons, protons, electrons, etc.) are conceived to be replaced by their corresponding antiparticles (antineutrons, antiprotons, positrons, etc.). Normal matter and antimatter would mutually annihilate each other upon contact and be converted into γ-rays.

Antineutrino. The antiparticle (see Antimatter) of the neutrino (q.v.).

Atomic mass units and atomic weight units. Before 1959 the unit of *atomic weight* (a.w.u.) was defined as

$$1 \text{ a.w.u.} = \tfrac{1}{16} \text{ of average mass of an oxygen atom in the natural mixture of isotopes (oxygen scale) of which the atomic weight is } 16 \cdot 0000$$

The unit of *nuclear or isotopic mass* (a.m.u. or *atomic mass unit*) was defined as

$$1 \text{ a.m.u. (O-16 scale)} = \tfrac{1}{16} \text{ of the mass of the O-16 isotope for which the isotopic mass} = 16 \cdot 0000$$
$$= \tfrac{1}{16}(16/N) = 1 \cdot 659\,81 \times 10^{-24} \text{ g}$$

Glossary of Nuclear Terms

where N is the Avogadro number $= 6\cdot 0248 \times 10^{23}$ atoms/mole
$=$ number of atoms in 1 mole of any isotope
$=$ number of atoms in 16 g of O-16
$=$ number of atoms in 12·0037 g of C-12

In 1959 the International Union of Pure and Applied Chemistry recommended that nuclear or isotopic masses and atomic weights should be based on the same scale, the unit of which is now known as the *unified atomic mass unit*, symbol, u. This is defined as

1 u (C-12 scale) $= \frac{1}{12}$ of the mass of the C-12 isotope of which the isotopic mass is 12·0000
$= \frac{1}{12}(12/N) = 1\cdot 660\ 53 \times 10^{-24}$ g

where N is the Avogadro number $= 6\cdot 022\ 169 \times 10^{23}$ atoms/mole
$=$ number of atoms in 1 mole of any isotope
$=$ number of atoms in 12 g of C-12
$=$ number of atoms in 15·9949 g of O-16

1 u $= 1\cdot 660\ 53 \times 10^{-24}$ g $= 1\cdot 000\ 434$ a.m.u. (O-16 scale)
$= 1\cdot 000\ 037$ a.w.u.

Atomic number, Z. The number of protons in the nucleus of an atom.

Atomic weight. The average weight of the atoms of an element. Various units of mass have been used for atomic weight scales (see above).

Autoradiography. A technique whereby the radioactive material in a specimen makes a radiograph of itself when placed in close contact with a photographic emulsion.

Background radiation. The radiation of the natural environment, consisting of that which comes from cosmic rays and from the naturally radioactive elements of the earth, including that from within the body.

Barn. A unit of area used in measuring the cross-sections of nuclei.
1 barn $= 10^{-24}$ cm^2
The term is a measure of the probability that a given nuclear reaction will occur.

Beta particle (β). An elementary particle emitted from a nucleus during radioactive decay. It has a single negative electric charge and a mass $\frac{1}{1837}$ of that of a proton. Beta particles are easily stopped by a thin sheet of aluminium or Perspex. A nucleus emitting a β-particle is transferred to an isotope of the element one place higher in the periodic table.

Binding energy. The binding energy of a nucleus is the minimum energy required to dissociate it completely into its component neutrons and protons. Neutron or proton binding energies are those required to remove a neutron or a proton, respectively, from a nucleus. Electron binding energy is that required to remove an electron completely from an atom or a molecule.

Body burden. The amount of radioactive material present in the body of a man or an animal.

Bremsstrahlung. Electromagnetic radiation emitted by charged particles when they are slowed down by electric fields in their passage through matter. Literally "braking radiation" (German).

Carrier. Inactive form of a radioactive substance that is added to give sufficient bulk to follow a chemical or physical process.

Carrier free. Radioactive material totally free from any inactive form of the material.

Castle. A housing for a counter and the radioactive material for assay that maintains a reproducible geometry between source and counter. A lead castle is used to reduce background radiation.

Compton effect. The interaction of a γ-photon with an outer electron in an atom. The photon is scattered with a longer wavelength and the electron is ejected from the atom, the energy being partitioned between the scattered photon and the electron.

Cosmic rays. Radiation of many sorts, but predominantly high-energy charged particles, that originates directly or indirectly from sources outside the earth's atmosphere. Cosmic radiation is part of the natural background radiation, and some of its constituents have extremely high energies (10^{12} eV).

Cross-section. A measure of the probability that a nuclear reaction will occur. Usually measured in barns (q.v.), it is the apparent area presented by a target nucleus (or particle) to an oncoming particle.

Curie (Ci). The unit of activity (q.v.).

Cyclotron. A machine that by means of an alternating electric field and a unidirectional magnetic field accelerates charged particles in a spiral path. Used mainly for isotope production and studies in high-energy nuclear physics.

Deuteron. The nucleus of an atom of deuterium, the heavy isotope of hydrogen, consisting of a neutron and a proton. Often used as the charged particle in the cyclotron (q.v.).

Dose, absorbed. The amount of ionizing radiation energy absorbed per unit mass of irradiated material at a specific location, such as a part of the human body. Measured in rads (q.v.).

Dose equivalent (D.E.). The product of the absorbed dose and the quality factor (q.v.) of the radiation. The unit of this dose is the rem (q.v.).

Dose, exposure. A measure of the intensity of a radiation field, expressed as the quantity of electric charge caused by the radiation in air. The unit is the röntgen (q.v.).

Dose, maximum permissible (M.P.D.). The dose of ionizing radiation which competent authorities have established as the maximum that can be received without undue risk to human health. Various limits are specified for different sections of the population.

Dose rate. The radiation dose delivered per unit time and measured, for instance, in rems per hour.

Effects of radiation, genetic. Effects that produce changes in those cells which give rise to egg or sperm cells and therefore affect offspring of the exposed individuals.

Effects of radiation, somatic. Effects limited to the general body cells of the exposed individual, as distinct from genetic effects. Large radiation doses can be fatal; smaller doses may make the individual noticeably ill or may produce temporary changes in blood-cell levels.

Electron. An elementary particle with a negative electric charge and a mass $\frac{1}{1837}$ that of the proton. Electrons surround the atom's positively charged nucleus and determine the atom's chemical properties. The charge on the electron is $1 \cdot 602 \times 10^{-19}$ C.

Electronvolt (eV). The kinetic energy gained by an electron when it has been accelerated through a potential difference of 1 V.

$$1 \text{ eV} = 1 \cdot 602 \times 10^{-19} \text{ J.}$$

Equivalence of mass and energy. In nuclear reactions mass and energy are interconvertible according to the relation

$$E = mc^2$$
$$E \text{ (joules)} = m \text{ (kg)} \times (2 \cdot 999 \, 9 \times 10^8)^2$$
$$E \text{ (MeV)} = m \text{ (u)} \times 931$$
$$1 \text{ u} \equiv 931 \text{ MeV}$$

Excited state. The state of an atom or nucleus when it possesses more than its normal energy. The excess energy is usually released as a photon, probably a visible or ultra-violet photon from an atom and a γ-ray photon from the nucleus.

Fission. The splitting of a heavy nucleus into two roughly equal parts (which are nuclei of lighter elements), accompanied by the release of a relatively large amount of energy and frequently one or more neutrons. Fission can occur spontaneously but usually is caused by the absorption of γ-rays, neutrons or other particles.

Fission products. The nuclei formed by the fission of heavy elements. They are of medium atomic weight, and almost all are radioactive. Examples are strontium-90, caesium-137 and iodine-131.

Fusion. The formation of a heavier nucleus from two lighter ones with the attendant release of energy, e.g.

$$^2_1\text{H} + ^2_1\text{H} \rightarrow ^3_2\text{He} + ^1_0\text{n} + 3 \cdot 3 \text{ MeV}$$
$$^2_1\text{H} + ^3_1\text{H} \rightarrow ^4_2\text{He} + ^1_0\text{n} + 17 \cdot 6 \text{ MeV}$$

Gamma-rays (γ). High-energy short-wavelength electromagnetic radiation emitted by a nucleus. Energies of γ-rays are usually between 0·010 and 10 MeV. X-rays also occur in this energy range, but are of non-nuclear origin, arising in the extra-nuclear electron cloud. Gamma radiation usually accompanies α- and β-emissions. Gamma rays are very penetrating and are best attenuated by dense materials like lead and concrete.

Half-life, biological (T_b). The time required for a biological system to eliminate, by natural processes, half the amount of a substance which has entered it.

Half-life, effective (T_{eff}). The time required for half the quantity of a radioactive substance present in a biological system to be eliminated from it by a combination of radioactivity decay and the biological elimination.

$$\frac{1}{T_{eff}} = \frac{1}{T_{1/2}} + \frac{1}{T_b}$$

where $T_{1/2}$ = radioactive half-life (q.v.).

Half-life, radioactive ($T_{1/2}$). The time in which half the atoms in a radioactive substance disintegrate.

Half-thickness layer. The thickness of an absorber that will reduce the radiation intensity from a radioactive material to one-half the original.

Ion. A charged atom or molecule.

Inverse-square law. $I_d = k/d^2$, where I_d is the intensity of the radiation at a distance d from the radioactive source, and k is a constant.

Isotopes. Atoms with the same atomic number (same chemical element) but different mass numbers. An equivalent statement is that the nuclei have the same number of protons but different numbers of neutrons. Thus $^{12}_{6}C$, $^{13}_{6}C$ and $^{14}_{6}C$ are isotopes of the element carbon.

K-electron capture (E.C.). The capture by an atom's nucleus of an orbital electron from the innermost or K-shell of electrons surrounding the nucleus. The atom is transformed to an isotope of the element one place lower in the periodic table.

Linear energy transfer (L.E.T.). A measure of the average energy lost in an absorbing medium per unit path length of the radiation. Its unit is the kiloelectronvolt per micrometre. Some examples are

	L.E.T. for water keV/μm
^{210}Po α-particles	136
^{3}H β-particles	3·2
^{60}Co γ-rays	0·3

Mass number (A). The sum of the numbers of neutrons and protons in a nucleus. The mass number of uranium-235 is 235. It is the nearest whole number to the atom's actual atomic weight.

MeV. Mega-electronvolt.

$$1 \text{ MeV} = 1·602 \times 10^{-13} \text{ J}$$

Neutrino. An electrically neutral elementary particle with a mass so small that it is extremely difficult to detect. It is produced in many nuclear reactions, for example in β-decay, and has high penetrating power. Neutrinos from the sun usually pass right through, or deep into, the earth.

Neutron. An elementary particle with no electric charge and a mass of 1·008 665 4 u. Neutrons involved in nuclear reactions are classified according to their kinetic energy as follows:

Thermal neutrons. Those having velocities comparable to the velocities due to thermal motion of the nuclei in surrounding material, namely 2200 m/s at 293 K, corresponding to a neutron kinetic energy of 0·025 eV.

Glossary of Nuclear Terms

Slow neutrons. Those having kinetic energies between 0·01 eV and a few electronvolts.

Epithermal (intermediate) neutrons. Those with kinetic energies intermediate between slow and fast neutrons.

Fast neutrons. Those with kinetic energies greater than 0·1 MeV.

Neutron capture. The reaction that occurs when an atomic nucleus absorbs or captures a neutron. The probability that a given material will absorb neutrons is proportional to its neutron capture cross-section and depends on the energy of the neutrons and the nature of the material.

Neutron flux. The intensity of neutron radiation expressed as the number of neutrons passing through 1 cm² in 1 s.

Neutron number, (N). The number of neutrons in the nucleus of an atom.

Nucleon. A constituent of the atomic nucleus; i.e. a proton or a neutron.

Nucleus. The small, positively charged core of an atom. Its diameter is only about $\frac{1}{10\,000}$ of that of the atom but it contains nearly all the mass. Except for hydrogen-1, all nuclei contain both protons and neutrons.

Nuclide. Any species of atom that exists for a measurable length of time. A nuclide can be distinguished by its atomic weight, atomic number and energy state. The term is used synonymously with isotope.

Pair production. The transformation of a high-energy γ-ray into a pair of particles (an electron and a positron) during its passage through matter. The γ-ray must have an energy which is at least equivalent to the rest mass of two electrons, i.e. 1·02 MeV. The process cannot occur for γ-rays of less energy than this and becomes more important for γ-rays of higher energies.

Photoelectric effect. The interaction of a γ-photon with a bound electron (usually a K-shell electron) in an atom. The γ-photon disappears and the electron is ejected with a kinetic energy equal to the energy of the photon less the binding energy of the electron. The process is an important absorption mechanism for the lower-energy γ-rays.

Photon. A discrete quantity of electromagnetic energy. Photons have momentum but no mass or electrical charge.

π-meson, or pion. An elementary particle, which may be neutral or charged. The mass of a charged pion is about 273 times that of an electron. An electrically neutral pion has a mass 264 times that of an electron.

Positron. An elementary particle with the same mass as an electron but positively charged. It is the antielectron (see Antimatter). It is emitted in some radioactive disintegrations and is produced in pair production. After losing its kinetic energy it interacts with an electron with subsequent annihilation (q.v.).

Proton. An elementary particle with a positive electric charge numerically equal to that of an electron and a mass of 1·007 276 6 u. The atomic number of an atom is equal to the number of protons in its nucleus.

Quality factor (Q.F.). A measure of the effectiveness of radiation to produce biological damage. Alpha particles have a Q.F. of 10 and γ-rays a Q.F. of 1, meaning that α-particles are ten times more effective in producing biological damage than γ-rays.

Quantum. The minimum discrete quantity of a physical parameter, such as energy or momentum, which is observed to be non-continuous in nature.

Rad. Radiation absorbed dose: the basic unit of absorbed dose of ionizing radiation. One rad is equal to the absorption of energy equal to $\frac{1}{100}$ J/kg of matter due to the interaction of the radiation.

Radiation protection. See entries under, Röntgen, Rem, Rad, Quality factor, Dose equivalent.

Radioisotope, or radio nuclide. An unstable isotope of an element whose nucleus decays or disintegrates spontaneously, emitting radiation.

Radiolysis. The alteration of molecules by radiation. Water subjected to irradiation forms hydrogen, oxygen and hydrogen peroxide.

Range. The maximum distance of penetration of radiation in an absorber.

Rem. Röntgen equivalent man. A unit of absorbed radiation dose in biological matter. It is equal to the absorbed dose in rads multiplied by the quality factor of the radiation. The unit is used only when discussing maximum permissible levels.

Röntgen (R). A unit of exposure dose (q.v.) of ionizing radiation.

$$1 R = 2\cdot 58 \times 10^{-4} \text{ C/kg}$$

Self-absorption. The absorption of the radiation from a labelled compound or solution in the main bulk of the compound or solution—a critical problem when dealing with weak β-emitters such as carbon-14 and sulphur-35.

Specific activity. The ratio of the activity of a labelled substance to the mass of the substance present. It is expressed as Ci/g or mCi/mM.

Specific gamma-ray constant (Γ). The exposure dose rate at a distance from a γ-emitting nuclide. The unit is the röntgen per millicurie per hour at 1 cm, or taking into account the inverse square law (q.v.), R mCi^{-1} h^{-1} cm^2.

Spontaneous fission. Fission that occurs without an external stimulus. Several heavy isotopes decay in this manner. Examples are curium-242, californium-252.

Stable isotope. A nuclide that does not undergo radioactive decay.

Tracer. An element or compound that has been made radioactive so that it can be easily followed (traced) in any reactions of the element. Radiation emitted by the radioisotope pinpoints its location. Stable isotopes can also be used as tracers, their presence being detected by mass spectrometry.

X-ray. Penetrating electromagnetic radiation emitted when the inner orbital electrons of an atom are excited and release energy; thus the radiation is non-nuclear. It is generated in practice by bombarding a metallic target with high-speed electrons.

Appendix 2

The thorium decay series

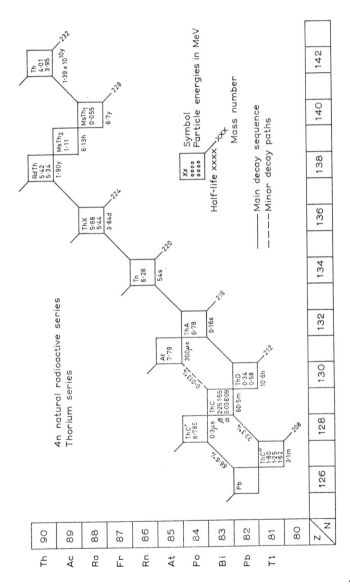

Appendix 3: The uranium decay series

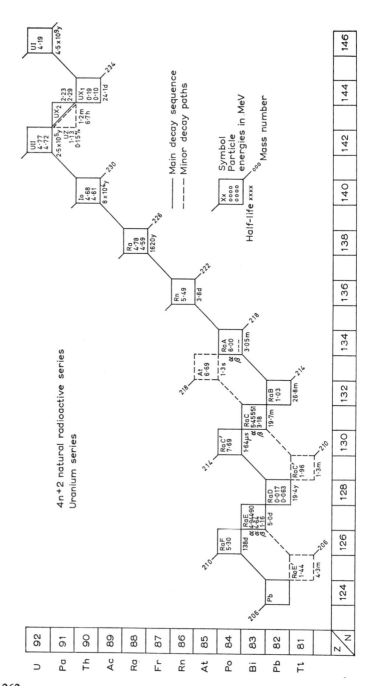

Appendix 4

Radioactivity data for uranium and thorium

Isotopic abundances, isotopic masses and atomic weights on the carbon-12 scale

Isotope	Abundance, atoms %	Isotope mass or atomic weight
Thorium-232	100·0	232·038 211 ± 0·000 042
Uranium-234	0·005 5 ± 0·000 2	234·040 90 ± 0·000 06
Uranium-235	0·720 4 ± 0·000 7	235·043 933 ± 0·000 043
Uranium-238	99·274 1 ± 0·000 7	238·050 76 ± 0·000 08
Natural uranium		238·03

Radioactive decay constants and half-lives

Isotope	Decay constant, year^{-1}	Half-life, $t_{1/2}$, years
Thorium-232	4·98 × 10^{-9}	1·39 × 10^{10}
Uranium-234	2·76 × 10^{-6}	2·48 × 10^{5}
Uranium-235	9·72 × 10^{-10}	7·13 × 10^{8}
Uranium-238	1·537 × 10^{-10}	4·51 × 10^{9}

Specific activities (in disintegrations per minute per milligram, dpm/mg)

Isotope	Specific activity of separated isotope, dpm/mg	Specific activity in natural uranium, dpm/mg	Specific activity in natural U_3O_8, dpm/mg
Thorium-232	246·0	—	—
Uranium-234	1·37 × 10^{7}	733·6 ± 1·6	622·1 ± 1·4
Uranium-235	4·74 × 10^{3}	33·7 ± 0·7	28·5 ± 0·6
Uranium-238	739·5	733·6 ± 1·6	622·1 ± 1·4

Specific activity of natural uranium and thorium

Natural thorium   0·111 µCi/g due to thorium-232 but excluding daughter nuclides

Natural uranium   0·33 µCi/g due to uranium-238 but excluding daughter nuclides
0·015 µCi/g due to uranium-235 but excluding daughter nuclides

Specific α-activity of natural uranium
1501 dpm/mg due to ^{238}U + ^{234}U + ^{235}U

Specific β-activity of natural uranium
733·6 dpm/mg due to ^{234m}Pa

Appendix 5

Table of beta/gamma emitting isotopes

Nuclide	Energy (MeV) and abundance of emissions			Specific γ-ray constant* (Γ)
	$t_{1/2}$	β	γ	
Caesium-137	30 y	0·51 95% 1·17 5%	0·662 from Ba-137m	3·3
Calcium-45	165 d	0·254 100%		
Carbon-14	5760 y	0·159 100%		
Cerium-144	285 d	0·19 19·5% 0·24 4·5% 0·32 76% see Pr-daughter	0·08–0·133	
Chlorine-36	3 × 10⁵ y	0·714 98·3% E.C 1·7%		
Cobalt-60	5·26 y	0·31 100%	1·17 100% 1·33 100%	13·2
Hydrogen-3 (tritium)	12·26 y	0·018		
Iodine-131	8·04 d	0·61 87% others 0·25–0·81	0·08–0·72	2·2
Lead-210 (radium-D)	22 y	0·017 85% 0·063 15%	0·047 4·1%	
Manganese-56	2·58 h	2·86 60% others 0·33–2·86	0·845 99% others 0·845–3·39	8·3
Phosphorus-32	14·3 d	1·71 100%		
Potassium-40	1·3 × 10⁹ y	1·32 89% E.C. 11%	1·46 11%	
Potassium-42	12·4 h	2·0 18% 3·6 82%	1·52 18%	1·4
Praseodymium-144	17 m	0·8–2·98 2·98 98%	0·69 1·6%	
Protoactinium-234 (UX₂)	1·18 m	0·58–2·31 2·31 90%	0·75, 1·00 others	
Silver-110 m	253 d	0·85 65% 0·53 33%	many 0·45–1·56	14·3
Sodium-22	2·6 y	β+0·54 90·5% E.C. 9·5%	0·51 from β+ 1·28 100%	12·0
Sodium-24	15·0 h	1·39 100%	1·37 100% 2·75 100%	18·4
Strontium-90	28 y	0·54 100% see Y-daughter		
Sulphur-35	87·2 d	0·167 100%		
Yttrium-90	64·2 h	2·27 100%		
Zinc-65	245 d	β+0·325 1·7% E.C. 98·3%	0·51 from β+ 1·11 49%	2·7

* In R mCi⁻¹ h⁻¹ cm² or mR μCi⁻¹ h⁻¹ cm².

Appendix 6

Lost counts corrections for Geiger-Müller counters

If the paralysis time or dead time of the counter is known, the percentage correction to be added to the observed count rate may be read from the appropriate graph.

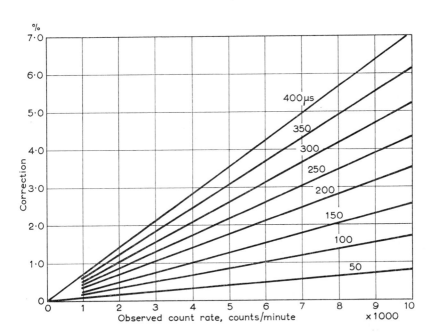

Appendix 7

Lost counts correction for a paralysis time of 400 μs

m	+0	+100	+200	+300	+400	+500	+600	+700	+800	+900
0	0	0	0	1	1	2	2	3	4	5
1 000	7	8	10	11	13	15	17	19	22	24
2 000	27	30	33	36	39	42	46	49	53	57
3 000	61	65	70	74	79	84	89	94	99	104
4 000	110	115	121	127	133	139	146	152	159	165
5 000	172	180	187	194	202	209	217	225	233	242
6 000	250	259	267	276	285	294	304	313	323	333
7 000	343	353	363	373	384	395	406	417	428	439
8 000	451	462	474	486	498	511	523	536	548	561
9 000	574	588	601	615	628	642	656	671	685	700
10 000	714	729	744	759	775	790	806	822	838	854
11 000	871	887	904	921	938	955	972	990	1008	1025
12 000	1043	1062	1080	1099	1117	1136	1155	1175	1194	1214
13 000	1234	1254	1274	1294	1314	1335	1356	1377	1398	1420
14 000	1441	1463	1485	1507	1529	1552	1574	1597	1620	1643
15 000	1667	1690	1714	1738	1762	1786	1811	1835	1860	1885
16 000	1910	1936	1961	1987	2013	2039	2066	2092	2119	2146
17 000	2173	2200	2228	2255	2283	2311	2340	2368	2397	2426
18 000	2455	2484	2513	2543	2573	2603	2633	2663	2694	2725
19 000	2756	2787	2818	2850	2882	2914	2946	2978	3011	3044
20 000	3077	3110	3144	3177	3211	3245	3279	3314	3349	3384
21 000	3419	3454	3489	3525	3561	3597	3634	3670	3707	3744
22 000	3781	3819	3856	3894	3932	3971	4009	4048	4087	4126
23 000	4165	4205	4245	4285	4325	4366	4406	4447	4488	4530
24 000	4571	4613	4655	4698	4740	4783	4826	4869	4912	4956
25 000	5000	5044	5088	5133	5178	5223	5268	5314	5359	5405
26 000	5452	5498	5545	5592	5639	5686	5734	5782	5830	5878
27 000	5927	5976	6025	6074	6124	6173	6224	6274	6324	6375
28 000	6426	6478	6529	6581	6633	6685	6738	6791	6844	6897
29 000	6950	7004	7058	7113	7167	7222	7277	7332	7388	7444

$n = m + nm\tau$

where n is the corrected counting rate, m is the observed counting rate and τ is the paralysis time. The quantities to be added to the observed counting rate, m (expressed in counts/minute), are tabulated for a paralysis time of 400 μs.

Appendix 8

Decay corrections

For radioactive decay the activity after-time t, $\left(\dfrac{dN}{dt}\right)_t$, is related to the activity at zero time, $\left(\dfrac{dN}{dt}\right)_0$, by the equation

$$\left(\frac{dN}{dt}\right)_t = \left(\frac{dN}{dt}\right)_0 e^{-\lambda t} = \left(\frac{dN}{dt}\right)_0 f$$

where the correction factor $f = e^{-\lambda t} = e^{-0.693\,147\,t/t_{1/2}}$ where $t_{1/2}$ is the half-life of the radioisotope involved.

Values of f are tabulated below against values of $t/t_{1/2}$. To make a decay correction, calculate the value of $t/t_{1/2}$ and read the corresponding value of f from the table.

f	$t/t_{1/2}$	f	$t/t_{1/2}$	f	$t/t_{1/2}$	f	$t/t_{1/2}$
0·99	0·014	0·74	0·434	0·49	1·029	0·24	2·059
0·98	0·029	0·73	0·454	0·48	1·059	0·23	2·120
0·97	0·044	0·72	0·474	0·47	1·089	0·22	2·184
0·96	0·059	0·71	0·494	0·46	1·120	0·21	2·252
0·95	0·074	0·70	0·515	0·45	1·152	0·20	2·322
0·94	0·089	0·69	0·535	0·44	1·184	0·19	2·396
0·93	0·105	0·68	0·556	0·43	1·218	0·18	2·474
0·92	0·120	0·67	0·578	0·42	1·252	0·17	2·556
0·91	0·136	0·66	0·599	0·41	1·286	0·16	2·644
0·90	0·152	0·65	0·621	0·40	1·322	0·15	2·737
0·89	0·168	0·64	0·644	0·39	1·358	0·14	2·837
0·88	0·184	0·63	0·667	0·38	1·396	0·13	2·943
0·87	0·201	0·62	0·690	0·37	1·434	0·12	3·059
0·86	0·218	0·61	0·713	0·36	1·474	0·11	3·184
0·85	0·234	0·60	0·737	0·35	1·515	0·10	3·322
0·84	0·252	0·59	0·761	0·34	1·556	0·09	3·474
0·83	0·269	0·58	0·786	0·33	1·599	0·08	3·644
0·82	0·286	0·57	0·811	0·32	1·644	0·07	3·837
0·81	0·304	0·56	0·837	0·31	1·690	0·06	4·059
0·80	0·322	0·55	0·862	0·30	1·737	0·05	4·322
0·79	0·340	0·54	0·889	0·29	1·786	0·04	4·644
0·78	0·358	0·53	0·916	0·28	1·837	0·03	5·059
0·77	0·377	0·52	0·943	0·27	1·889	0·02	5·644
0·76	0·396	0·51	0·971	0·26	1·943	0·01	6·644
0·75	0·415	0·50	1·000	0·25	2·000		

Appendix 9

The electromagnetic spectrum

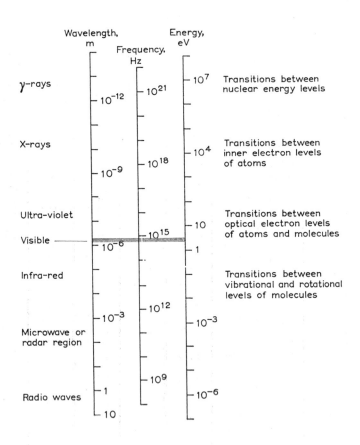

Appendix 10

Backscatter peak and Compton edge energies

(see next page)

The incident and scattered energies of a γ-ray photon which undergoes Compton scattering through an angle θ are related thus:

$$E_{\gamma\,scattered} = E_{\gamma\,incident} \times \frac{1}{1 + \alpha(1 - \cos\theta)}$$

where $\alpha = E_{\gamma\,incident}/m_0 c^2$.

The minimum value of $E_{\gamma\,scattered}$ occurs when $\theta = 180°$ and is given by

$$E_{\gamma\,scattered} = E_{\gamma\,incident} \times \frac{1}{1 + 2\alpha}$$

The backscatter peak occurs just above $E_{\gamma\,scattered(minimum)}$ and the Compton edge occurs at $E_{\gamma\,incident} - E_{\gamma\,scattered(minimum)}$.

$E_{\gamma\,incident}$	$E_{\gamma\,scattered(min)}$	$E_{\gamma\,incident} - E_{\gamma\,scattered(min)}$
0·01	0·010	0·000
0·02	0·019	0·001
0·03	0·027	0·003
0·04	0·035	0·005
0·05	0·042	0·008
0·06	0·049	0·011
0·07	0·055	0·015
0·08	0·061	0·019
0·09	0·067	0·023
0·10	0·072	0·028
0·15	0·095	0·055
0·20	0·112	0·088
0·25	0·126	0·124
0·30	0·138	0·162
0·35	0·148	0·202
0·40	0·156	0·244
0·45	0·163	0·287
0·50	0·169	0·331
0·55	0·174	0·376
0·60	0·179	0·421
0·65	0·183	0·467
0·70	0·187	0·513
0·75	0·191	0·559
0·80	0·194	0·606
0·85	0·196	0·654
0·90	0·199	0·701
0·95	0·201	0·749
1·00	0·204	0·796
1·10	0·207	0·893
1·20	0·211	0·989
1·30	0·214	1·086
1·40	0·216	1·184
1·50	0·218	1·282
1·60	0·220	1·380
1·70	0·222	1·478
1·80	0·224	1·576
1·90	0·225	1·675
2·00	0·227	1·773
3·00	0·235	2·765
4·00	0·240	3·760
5·00	0·243	4·757
6·00	0·245	5·755
7·00	0·247	6·753
8·00	0·248	7·752
9·00	0·248	8·752
10·00	0·249	9·751

Appendix 11

Some physical constants

Rest mass of proton	M_p	1.6725×10^{-27} kg
Rest mass of neutron	M_n	1.6748×10^{-27} kg
Rest mass of α-particle	M	6.6442×10^{-27} kg
Rest mass of electron	m_0	9.1091×10^{-31} kg
Electronic charge	e	1.6021×10^{-20} e.m.u.
		$= 1.6021 \times 10^{-19}$ C
		$= 4.80296 \times 10^{-10}$ e.s.u.
Charge/mass ratio for electron	e/m	1.7588×10^{11} C/kg
		$= 5.2727 \times 10^{17}$ e.s.u./g
Charge/mass ratio for α-particle		4.8223×10^{7} C/kg
		$= 1.4457 \times 10^{14}$ e.s.u./g
Atomic mass unit	u	1.66053×10^{-27} kg
Mass/energy equivalence		1 u = 931 MeV
Energy equivalent of electron rest mass		0.511 MeV
Electronvolt	eV	1.6021×10^{-19} J
		$= 1.6021 \times 10^{-12}$ ergs
Speed of light in vacuum	c	2.997925×10^{8} m/s
Avogadro number		6.022169×10^{23} atoms/mole

Appendix II

Useful Addresses

Index

Absorbed dose, 75, 77, 256
Absorption
  of α-particles, 34, 135
  of β-particles, 141
  of γ-rays, 159
Activity, 103
Alpha decay, 10
  energy release in, 12
Alpha particles, 1, 12, 131, 254
  absorption of, 34
  absorption in air, 132
  absorption in aluminium, 133, 135
  deflection of, in magnetic field, 149
  energy of, 131
  ion pairs formed by, 117
  magnetic rigidity of, 149, 151
  range of, 131, 133, 138
  scattering of, 2, 152
  straggling, 131
Alpha recoil, 196
Annihilation γ-rays, 35, 167, 254
Antimatter, 254
Antineutrino, 10, 13, 14, 141, 254
Apparent specific activity, 175
Artificial radioisotopes, 20
Atomic mass scale, 4, 255
Atomic mass unit, 254
Atomic number, 2, 255
Atomic weight, 2, 255
Attenuation coefficient,
  linear, 159
  mass, 160, 161
Autoradiography, 26, 28, 225, 243
  artefacts in, 227, 244
  exposure times for, 227
  macroscopic, 243
  microscopic, 249
  stripping film technique, 249

Background radiation, 62, 99, 255
Backscatter peak, 169, 269
Band theory of solids, 48
Barn, 22, 255
Beta decay, 10
  energy release in, 13
Beta particles, 1, 4, 255
  absorbed dose from, 86
  absorption of, 35, 141
  deflection in magnetic field, 147
  energy determination, 140
  ion pair formation by, 118
  magnetic rigidity of, 147

range/energy relationship, 142
relativistic mass, 140
spectra, 14, 141
Binding energy, 5, 255
Biological cell, 63
Biosynthesis of labelled compounds, 28
Bismuth-212 separation, 195
Body burden, 78, 256
  maximum permissible, 80
Bremsstrahlung, 141, 256

Capture gamma, 21
Carbon fixation, 228, 232
Carbon-14-labelled compounds, 26
Carrier, 25, 91, 256
Carrier-free radioisotopes, 24, 256
Castle, 46, 47, 256
Centrosomes, 64
Chemical purity, 30
Chemical synthesis of labelled compounds, 27
Chi-squared test, 129
Chromatogram scanner, 220
Chromosomes, 63
Cloud chamber, 33, 164
Codon, 65
Compton edge, 169, 269
Compton effect, 256
Compton scattering, 35, 36, 167, 256
Conduction band, 49, 56
Containment, 88, 90
Coulomb barrier, 152
Coulomb repulsion forces, 20, 152
Counting percentage, 176
Cosmic rays, 256
Critical organs, 77
Cross-sections, 22, 202, 256
Curie, 23, 254, 256
Cyclotron, 256
  produced isotopes, 21
Cytoplasm, 63

Decay constant, 18, 103
Decay diagrams
  for α-emission, 12
  for β^--emission, 14
  for β^+-emission, 15
  for electron capture, 17
  for γ-ray emission, 18
Decay law, 18, 103
Decay, radioactive, 103
  correction for, 267

273

Decay (*Contd.*)
  simulation of, 109
  statistics of, 124
Decay rate, 25
Dead time, 43, 122
Decontamination, 95
Demountable filter, 94
Deoxyribonucleic acid (DNA), 64, 69
Depletion layer, 51
D.E.S. regulations, 98
Detector,
  efficiency, 172
  gas ionization, 36
  Geiger–Müller, 39
  ionization chamber, 39, 136
  pulse size, 37
  scintillation, 52
  spark counter, 38, 48, 137
Deuteron, 21, 256
Developer, Kodak D19b, 209, 228
Disintegration energy, 12
  for α-particle, 12
  for β^--particle, 13, 14
  for β^+-particle, 15, 16
  for K-electron capture, 16, 17
Dominant character, 20
Dose,
  absorbed, 75, 77, 256
  equivalent, 77, 256
  exposure, 74, 256
  maximum permissible, 256
  radiation, 74
Dose rate, 256
  of approved sources, 100
Dose-rate meter, 85, 99
Dosimeter, for α-particle detection, 136
Double containment, 90
Double helix, 65, 66
Dynode, 58

Effects of ionizing radiation, 72
Efficiency of detectors, 121, 172
Electromagnetic spectrum, 268
Electron, 2, 140, 257
  avalanche, 40
  multiplication, 59
  -positron pair, 36, 159, 167
Electronvolt, 6, 257
Electroscope, 115, 136, 189
Energy/mass relationship, 5, 12, 35, 257
Enzyme, 68
Errors in counting rate,
  combinations of, 128
  non-random, 129, 130
  random, 125, 130
Escape peaks, 167
Excitation, 39, 40, 257
Exciton, 55
Exposure dose, 74, 256
  calculation of, 85

Feather equation, 142
Fission, 7, 257
  products, 257
Fluorescence, 52
Forbidden bands, 49, 53, 56
Fusion, 7, 20, 257

Gametes, 20
Gamma rays, 1, 18, 257
  absorption of, 35, 159
  exposure dose rate for, 85
Gamma-ray scintillation spectrometry, 167
Gamma-ray sources, 170
Gamma-ray spectrum, 61, 168
Gas amplification factor, 38, 120
Gaussian distribution law, 125
Geiger–Müller counter,
  breakdown voltage of, 120
  dead time of, 43, 122
  efficiency of, 121
  end-window, 45
  for γ-ray counting, 47
  instrumentation for, 47
  liquid sample, 47
  lost counts corrections, 45, 265, 266
  operating voltage, 120
  paralysis time, 43
  pulse shapes of, 42
  recovery time, 43
  starting voltage, 120
  threshold voltage, 120
Genes, 70
Genetic code, 65, 68
Genetic effects of radiation, 256
Genotype, 70
Geometry of counter, 46
Glendenin equation, 142
Golgi apparatus, 63
Growth factor, 23, 24

Half-life, 18, 103, 258
  analogy, 105
  biological, 78, 257
  determination of, 103
  effective, 78, 257

Index

Half-thickness, 161, 258
Hazards of radiation, 77

I.C.R.P., 78
Impact parameter, 153, 155, 157
Infinitely thick source, 174
Intrinsic efficiency, 170
Inverse square law, 88, 162, 258
Iodine-128 preparation, 202
Ion, 258
  movement in plants, 240, 241
  uptake in plants, 247
Ion pair, 34, 117
  energy of forming, 117
  onization, 34, 115
    by α-particles, 117, 131
    by β-particles, 118
  chamber, 38, 115
  chamber region, 37
  current, 39, 117
  potential, 39, 41
Isobars, 9, 11
Isotopes, 3, 258
Isotopic abundance, 22

K-electron capture, 11, 16, 17, 258

Labelled compounds, 29
Laboratory procedures,
  for closed sources, 89
  for open sources, 89
Laboratory regulations, 97
Lead-212 separation, 195
Linear absorption or attenuation coefficient, 159
Linear energy transfer, 76, 258
Lithium-drifted detector, 51
Luminescence, 52
Lysosomes, 64

Magnetic rigidity, 147, 151
Mass absorption or attenuation coefficient, 161, 179
Mass number, 3, 6, 258
Maximum permissible body burden, 80
Maximum permissible concentrations, 81, 101
Maximum permissible doses, 80
Maximum permissible levels, 73, 79
  in educational establishments, 98
Mega-electronvolt, 258
Mesons, 5, 259
Messenger ribonucleic acid (m-RNA), 65

Mineral nutrition, 240
Mitochondria, 63
Mutations, 69, 70

Neutrino, 11, 141, 258
Neutron, 3, 258
  capture, 20, 259
  flux, 22, 259
  number, 259
  sources, 202
Nuclear
  emulsions, 207
  fission, 22
  forces, 4
  reaction rate, 22
  stability, 4
  theory of Rutherford, 152
Nucleons, 5, 259
Nucleus, 2, 259
Nuclide, 9, 259
Nuclides chart, 8, 9

Optimum counting time, 129

Pair production, 35, 167, 259
Paralysis time, 43, 122
  corrections for, 105, 265, 266
Particle flux, 22, 23
Partition coefficient, 215
Phenotype, 70
Phosphors, 52
  table of, 54
Phosphorescence, 52
Photoelectric effect, 35, 159, 167, 259
Photofraction, 169
Photo-ionization, 39
Photomultiplier, 58
Photon, 259
Photopeak, 168
Photosynthesis, 224
Physical constants, 271
Pipettes for radiochemical work, 91
Planchettes, 46, 93
Plasmalemma, 240, 241
Plateau of Geiger–Müller counter, 120, 122
Poiseuille's equation, 107
Poisson distribution law, 126
Positron, 11, 35, 259
Potential energy well, 152
Praseodymium-144 separation, 197
Proportional counter region, 38
Protective clothing, 89, 90
Protoactinium-234m separation, 190
Proton, 3, 10, 259

Pulse
  amplification, 60
  analyser, 60
  electroscope, 115, 134, 188
  size, voltage characteristic, 37
Purity of labelled materials, 30
Quality factor, 76, 259
Quantum, 260
Quenching, 41
Rad, 76, 260
Radiation
  decomposition, 31
  interaction with matter, 33
  protection, 62, 86–97, 260
Radioactive decay, 10, 103
  statistics of, 124
  water analogue, 107
Radioactive equilibrium, 111
Radioactivity, detection of, 33
Radiochemical purity, 30
Radiocolloid, 31, 192
Radioisotopes, 260
  production, 26
  table of, 264
  toxicity classification of, 82
Radioisotopic purity, 30
Radiolysis, 260
Radiometric analysis, 217
  copper, 222
  lead, 219
  magnesium, 217
Radon-220 separation, 188
Range, 35, 260
  of α-particles, 131
  of β-particles, 140
  measurement of, for α-particles, 133
  measurement of, for β-particles, 140
  table of, for α-particles, 133
Range/energy relationship for β-particles, 143
Reaction cross-sections, 22
Recessive character, 70
Recoil energy, 204
Recovery time, 43
Relativistic mass, 140
Rem, 77, 260
Resolution,
  in γ-spectrometry, 169
  in autoradiography, 227, 244, 249
Respiration, 234
  in animals, 236
  in bacteria, 238
  in plants, 235
Rest mass of electron, 140, 167

Ribonucleic acid (RNA), 65
Ribosome, 63
Röntgen, 75, 260
Saturation factor, 24, 202
Scattering angle, 153, 269
Scintillation counter, 52
Scintillation spectrometry, 167
Secular equilibrium, 190, 195, 201, 210, 211
Self-absorption, 174–181, 260
Semiconductor detectors, 48, 50, 51
Shielding, 87
Sickle cell anaemia, 68
Solubility determination, 182
  of silver chloride, 183
  of strontium carbonate, 185
Somatic effects of radiation, 257
Source preparation, 93–95
Spark counter, 38, 48, 137
Specific activity, 24, 175, 260
Specific γ-ray constant, 85, 260
Specific ionization, 37
Specific surface area, 213
Spontaneous fission, 260
Stable isotope, 260
Standard deviation, 127
Standard man, 78, 81, 101
Starting voltage of Geiger–Müller counter, 120
Statistics of decay, 124
Straggling of α-particles, 131
Stripping film, 249
Summation peak, 169
Szilard–Chalmers reaction, 203
Thickness,
  absorber, 141, 161, 179
  linear, 159
Thorium decay,
  data, 263
  series, 261
Thorium stars, 211
Thorium-234 separation, carrier-free, 192
Thorium-X (radium-224) separation, 211
Threshold voltage of the Geiger–Müller counter, 120
Tonoplast, 240, 241
Total energy peak, 168
Toxicity of radioisotopes, 82
Tracer, 260
Track density, 207
Transfer ribonucleic acid (t-RNA), 67
Transient equilibrium, 112

Index

Translocation in plants, 230
  of ions, 241, 244
  of sugars, 235
Tritium-labelled compounds, 28, 30
True specific activity, 176

Unified atomic mass unit, 4, 255
Uranium decay,
  data, 263
  series, 262

UX_2 cow, 190

Valence band, 49, 56
Visual range of β-particles, 144

Waste disposal, 96, 102
Wavelength shifter, 58

X-rays, 257, 260
  from K-electron capture, 17, 33

Plate 1 End-window Geiger–Müller counter in a lead castle, with a quench unit and a scaler incorporating an e.h.v. supply

(*Nuclear Enterprises Ltd.*)

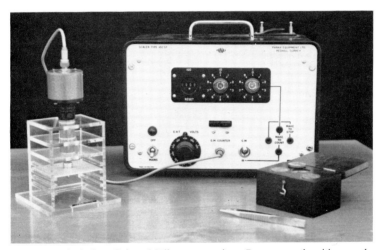

Plate 2 End-window Geiger–Müller counter in a Perspex castle with a scaler incorporating an e.h.v. supply

(*Philip Harris Ltd. and Panox Ltd.*)

Plate 3 Perspex castle with a Geiger–Müller counter for liquid samples and a chromatogram scanner

Plate 4 Solid-state detectors for α-particle counting

Plate 5 Laboratory bench prepared for work with open sources of radioisotopes

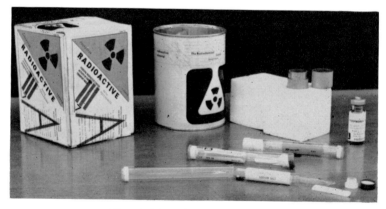

Plate 6 Examples of isotope packaging

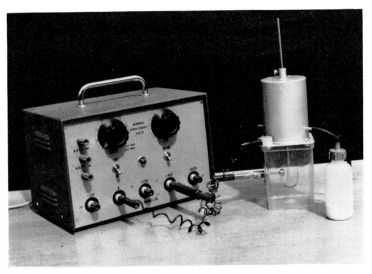

Plate 7 Pulse electroscope with a radon source bottle

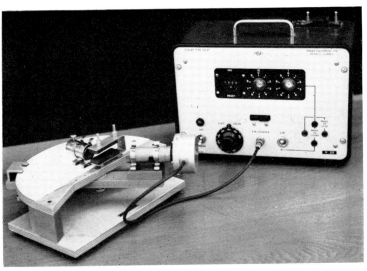

Plate 8 Apparatus for studying deflection of β-particles in a magnetic field
(*Philip Harris Ltd. and Panox Ltd.*)

Plate 9   Experimental analogy for studying the scattering of α-particles by heavy nuclei

(*Philip Harris Ltd.*)

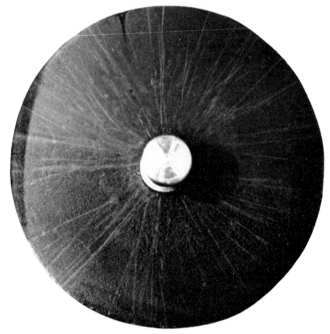

Plate 10   Tracks of α-particles in an expansion-type cloud chamber

Plate 11 Apparatus for the separation of lead-212 and bismuth-212 from the thorium series
(*Philip Harris Ltd.*)

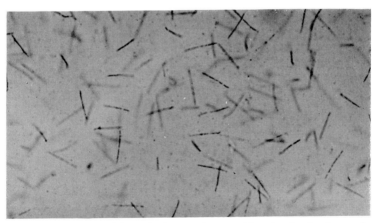

Plate 12 Alpha-particle tracks in a nuclear emulsion due to the decay of uranium-238 (4·2 MeV) and uranium-234 (4·77 MeV)

The nuclear emulsion plate was loaded with uranium by immersion in uranyl acetate solution
× 670

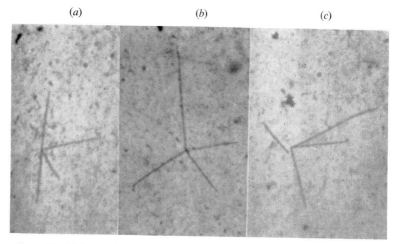

Plate 13 Alpha-particle tracks in a nuclear emulsion due to the decay of radium-224

a) A star with approximately equal track lengths due to decay through radium-224 (5·68 MeV), radon-220 (6·28 MeV), polonium-216 (6·78 MeV) and bismuth-212 (6·04 MeV)
b) A star with one markedly longer track due to decay through radium-224, radon-220, polonium-216 and polonium-212 (8·78 MeV)
c) A star which shows diffusion of the radon-220 atom in the emulsion before the decay sequence continued

× 670

Plate 14 Macroscopic autoradiograph of a leaf labelled with carbon-14 by photosynthesis

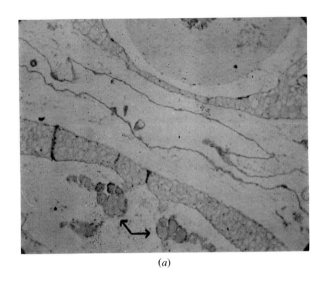

(a)

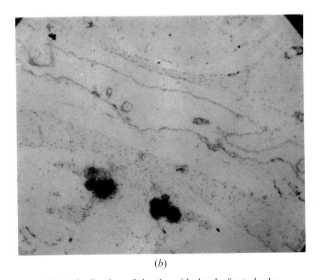

(b)

Plate 15 Section of the thyroid gland of a tadpole
(a) Photomicrograph showing the two lobes of the gland (arrowed)
(b) Autoradiograph showing heavy labelling by iodine-131
× 75

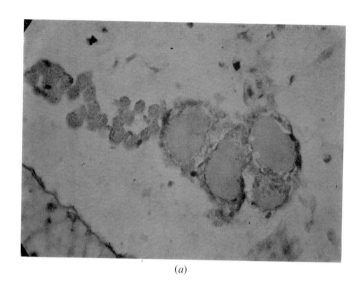

(a)

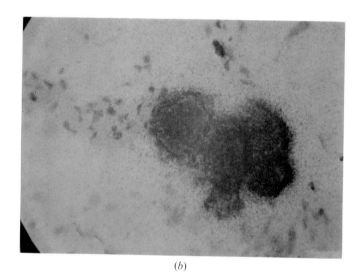

(b)

Plate 16 Right lobe of the thyroid gland shown in Plate 15 under higher magnification

(a) photomicrograph (b) autoradiograph

× 330